EVALUATING NOVEL THREATS TO THE HOMELAND

UNMANNED AERIAL VEHICLES AND CRUISE MISSILES

Brian A. Jackson, David R. Frelinger
Michael J. Lostumbo, Robert W. Button

Prepared for the Defense Threat Reduction Agency

RAND NATIONAL DEFENSE RESEARCH INSTITUTE

The research described in this report was prepared for the Defense Threat Reduction Agency. The research was conducted in the RAND National Defense Research Institute, a federally funded research and development center sponsored by the Office of the Secretary of Defense, the Joint Staff, the Unified Combatant Commands, the Department of the Navy, the Marine Corps, the defense agencies, and the defense Intelligence Community under Contract W74V8H-06-C-0002.

Library of Congress Cataloging-in-Publication Data is available for this publication.

ISBN 978-0-8330-4169-2

The RAND Corporation is a nonprofit research organization providing objective analysis and effective solutions that address the challenges facing the public and private sectors around the world. RAND's publications do not necessarily reflect the opinions of its research clients and sponsors.

Cover Design by Stephen Bloodsworth

Published 2008 by the RAND Corporation
1776 Main Street, P.O. Box 2138, Santa Monica, CA 90407-2138
1200 South Hayes Street, Arlington, VA 22202-5050
4570 Fifth Avenue, Suite 600, Pittsburgh, PA 15213-2665
RAND URL: http://www.rand.org/
To order RAND documents or to obtain additional information, contact
Distribution Services: Telephone: (310) 451-7002;
Fax: (310) 451-6915; Email: order@rand.org

Preface

Deciding how to invest homeland security resources wisely in the United States can often appear to be an intractable problem because the large, open American society seems to be so vulnerable to so many threats in every corner of the country. This monograph is intended to help bound the problem in order to aid policy and resource decisions about one type of potential threat to the homeland: cruise missiles and unmanned aerial vehicles (UAVs). Importantly, the methodology used can be applied to other modes of attack, and the insights gained from this methodology extend to other threats as well. The focus of the research is on a specific class of weapons, but those weapons are not assessed in isolation; rather, it considers class of weapons as one of many options open to a potential attacker and seeks to identify investment strategies that are effective against multiple threats.

This monograph should be of interest to homeland security policymakers, military and defense planners, analysts examining the terrorist threat, technology and defense system designers, and individuals charged with protecting potential targets in the U.S. homeland from terrorist attack.

This research was sponsored by the Defense Threat Reduction Agency (DTRA) and conducted within the International Security and Defense Policy Center of the RAND National Defense Research Institute, a federally funded research and development center sponsored by the Office of the Secretary of Defense, the Joint Staff, the Unified Combatant Commands, the Department of the Navy, the Marine Corps, the defense agencies, and the defense Intelligence Community.

For more information on RAND's International Security and Defense Policy Center, contact the Director, James Dobbins. He can be reached by email at Dobbins@rand.org; by phone at 703-413-1100, extension 5134; or by mail at RAND, 1200 South Hayes Street, Arlington, Virginia 22202. More information about RAND is available at www.rand.org.

Contents

Figures

Tables

Summary

How to invest homeland security resources wisely in the United States can appear to be an intractable problem because the large, open American society seems to be vulnerable to so many threats in every corner of the country. This monograph is intended to present a defense-planning approach to bound the problem and thereby aid policy and resource decisions about one type of potential threat to the homeland: cruise missiles and unmanned aerial vehicles (UAVs). The methodology used can be applied to other modes of attack, and the insights gained from this approach extend to other threats as well. Indeed, although the focus of the research is on a specific class of weapons, it does not look at that class in isolation; rather, it considers the weapons as one of many options open to a potential attacker and seeks to identify investment strategies that are effective against multiple threats and weapons.

Cruise missiles and UAVs are the chosen focus of this monograph because they represent important tools in the arsenal of the U.S. military. The U.S. military has demonstrated their utility in modern combat in many recent conflicts. Therefore, it should not be surprising that cruise missiles and UAVs are increasingly entering the inventories of militaries around the world, and even those of some terrorist organizations. Cruise missiles are at times dubbed "the poor man's air force": In some circumstances, they can achieve similar effects to that of fixed-winged aircraft for a fraction of the cost. And, although perhaps not as illustrious as ballistic missiles, cruise missiles carry a certain status for countries and militaries as a milestone in weapon prowess and technical advancement. But how much of a threat do these capabilities represent?

The difficulty in answering this question stems from intelligence and law-enforcement organizations' limited ability to monitor terrorist organizations and uncover new attack modes before they have been used in an attack. For instance, in July 2006, Hezbollah apparently surprised both Israeli and U.S. intelligence organizations when it attacked an Israeli naval vessel with a C-802 anti-ship cruise missile. That limited ability means that, in planning defenses, a traditional intelligence threat assessment cannot focus only on known or likely attack modes. Instead, defense planners must consider plausible attack modes, including weapons that could be transferred from a national military to a terrorist organization, particularly those that can be operated by a small number of people and do not require large infrastructure or support investments.

Assessing how such weapons could be used in attacks in the United States is also difficult, because there is also an almost infinite number of targets within the homeland that are vulnerable from the air and therefore represent possible sites for attack. For such a challenge as thinking about how to respond to the potential use of these weapons or the design of defensive approaches, an unbounded problem becomes intractable: The resource requirements of protecting everything quickly become staggering. This challenge is further complicated because such weapons represent only one from a variety of attack options an adversary could choose to use. Before the country invests in a wide array of cruise-missile or other air-defense assets for the nation, the problem needs to be bounded so that scarce resources can be focused productively.

Examining the Threat from UAVs and Cruise Missiles via a "Red Analysis of Alternatives"

In essence, to assess the threat of cruise missiles and UAVs to the homeland, we cannot consider them in isolation; instead, we must consider the problem from the attacker's point of view, in which these systems are only one of many ways to stage an attack. We call this approach a "red analysis of alternatives" because it will consider cruise missiles and UAVs as one option among many attack possibilities from the point of view of a potential adversary. As a result, we designed our analysis to explicitly compare these systems against other ways in which adversaries could choose to stage offensive operations and to explicitly test whether (and in what specific operational situations) UAVs and cruise missiles provided significant advantages over those alternatives.

The advantages provided by UAVs and cruise missiles over other attack modes are not in the destructive power that they can carry; they are in the way they carry that power and the distance from which they allow an adversary to control its delivery. The value of this advantage to an adversary and, as a result, the likely attractiveness of these systems will therefore be driven by the benefits of aerial attack in solving specific operational problems.

UAVs and cruise missiles are most likely to be attractive in situations in which their aerial, long-standoff capability solves key operational problems an attacker faces in planning and mounting an operation. These systems appear most advantageous because they could make it easier for an adversary to do five main things:

1. attack over perimeter defenses
2. attack over national borders
3. carry out multiple simultaneous attacks
4. conduct an attack campaign (a series of attacks over time)
5. attack area targets with unconventional weapons.

Looking specifically at how adversaries can perform these specific tasks enables analysis of UAVs and cruise missiles in the context of where their advantages are likely to be most important. Through such an analysis, it is possible to identify the key characteristics of the systems that distinguish them from other means of attack and highlight the specific factors that might lead adversaries to acquire and use them.

After analyzing cruise missiles and UAVs in their most favorable light from the attacker's perspective, we conclude that they do not appear to have major advantages over other ways of carrying out operations against similar targets, although they cannot be dismissed outright as a potential threat. Where they did appear preferable, the choice for these systems was driven by the actions of the defense or in-place security measures—i.e., were alternative attack modes foreclosed by defenses or did concerns about a potentially compromised plan push the attacking group farther away from its desired targets? The price of these advantages was, however, greater complexity, technological uncertainty, and higher cost and risks associated with these platforms. Consequently, rather than being an attack mode likely to be widely embraced by such actors, UAVs and cruise missiles appear to represent a "niche threat"—potentially making some contribution to the overall asymmetric and terrorist threat. Cruise missiles and UAVs do provide some advantages to an attacker, but in most cases there are simpler alternatives that provide similar, or even superior, capabilities.

Considering Defensive Approaches

In considering appropriate defensive responses to these systems, the homeland-security planner must weigh the scale of investments that are appropriate given the nature of the threat they pose. In view of the availability of alternative attack modes and the uncertainties associated with the success of cruise missiles and UAVs to adversaries, broad-based and expensive efforts focused only on this specific threat appear unrealistic. Given resource constraints, defense planning must therefore also include a broad examination of all the defensive options that are available to craft a prudent and realistic response.

Efforts to defend against this threat could be directed in a wide variety of ways, ranging from counterproliferation efforts to limit technology acquisition, to counterterrorism targeting groups' procuring the devices, to recovery plans for addressing the consequences of attacks if they do occur. From a comparison of the options and qualitative examination of their costs, we conclude that a prudent defensive strategy appropriate to the magnitude of the cruise-missile and UAV threat would focus primarily on counterterrorism and law enforcement to prevent attacks and on measures to mitigate the results of such attacks and quickly recover after they occur. Such an investment will increase security not only against cruise-missile and UAV attacks but against a wide variety of potential terrorist attacks.

Some modest defensive investments specific to cruise missiles and UAVs are called for. The Defense Department and the Intelligence Community should gather information to help law enforcement identify potential supply chains and conduct forensics analysis of these systems. Collection of relevant technology and information to support the development of better forensics approaches—e.g., acquisition and study of foreign UAV and cruise-missile systems in ongoing efforts to gather and exploit technical intelligence—has an important role in building the foundations needed for post-attack study and for determining any unique signatures of specific countries' systems. The key to gaining the full benefits of such activity is the ability to share relevant information with law-enforcement organizations.

In addition, diplomatic efforts to strengthen international arms control regimes, particularly those focused on long-range and large-payload air vehicles, could make it more difficult for adversaries to obtain the most destructive of these systems.

In our examination of defensive options, we assessed the potential for deploying active defenses to shoot down cruise missiles and UAVs; nevertheless, we do not recommend broadly investing in such defenses for use in the homeland. Relative to the threat posed by UAVs and cruise missiles, active defense systems are too costly to operate, can defend only very small areas, and have limitations even within these small, defended areas. It is our conclusion that investments in defenses at the point of attack will take away resources from other more-productive defense investments focused on preventing a much wider range of attacks before they occur.

Acknowledgments

We wish to thank our sponsor, the Advanced Systems and Concepts Office of the Defense Threat Reduction Agency. In particular, Jonathan D. Fox offered continuing support and interest throughout this project, and David Hamon had the foresight to see the need for this research. We also wish to thank our colleague David Mosher of RAND and Dennis Pluchinsky, a former senior terrorism analyst in the U.S. Department of State and currently with Transecur, Inc., for reviewing an earlier draft of this monograph. All shortcomings obviously remain our responsibility.

Abbreviations

CBR	chemical, biological, and radiological
CM	cruise missile
CT	counterterrorism
DoD	Department of Defense
DTRA	Defense Threat Reduction Agency
FAA	Federal Aviation Administration
FARC	Revolutionary Armed Forces of Colombia
GPS	Global Positioning System
ICBM	intercontinental ballistic missile
INS	Inertial Navigation System
ISR	intelligence, surveillance, and reconnaissance
kg	kilogram
km	kilometer
kt	knot
lb	pound
LE	law enforcement
m	meter

m^2	meters squared
mi	mile
MTCR	Missile Technology Control Regime
nmi	nautical mile
PIRA	Provisional Irish Republican Army
psi	pounds per square inch
UAS	unmanned aerial systems
UAV	unmanned aerial vehicle
WMD	weapon of mass destruction

CHAPTER ONE

Introduction

Adoption of new technology by adversaries—whether hostile states or violent nonstate groups—frequently requires that security planners assess and respond to novel threats. The development of nuclear weapons in World War II required major adjustments in thinking by military tacticians, because the shift in environment produced by the proliferation of those weapons changed the security landscape. The attacks of September 11, 2001, wherein the use of airliners as weapons to produce mass casualties shocked traditional views of the capabilities of nonstate organizations and the nature of the threat they posed, have similarly challenged established security concepts. In the wake of the shifts brought about by both these now-historic cases, major efforts were focused on understanding both what had happened and how the world had changed, and the reaction to the shifts led to redoubled efforts at foresight to better understand whether more such shifts were on the horizon, to prepare for them before they arrived.

Defense planners must assess how a shift in technology or tactics changes the balance in current or potential conflicts. Threats with the potential to be very disruptive may necessitate specific and focused responses to prevent or hedge against the effects of such disruption. Responding to each threat in isolation is not free, however. Even in wealthy nations, resources are finite and development of new responses to every threat that arises has the potential to spread a defensive effort thin or to dissipate resources that would be better used in pursuit of other national goals. Consequently, security planners must examine new threats to determine what about them is *not* novel. By exploring a new threat's similarities with dangers the United States already faces, analysts can determine whether the threats are covered by defensive efforts that are in place. Such a balanced approach aims both to identify threats that merit special attention and, by also identifying those that do not, to marshal resources that may be needed for other purposes.

Unmanned Aerial Vehicles and Cruise Missiles: Technological Change Producing a Potential Threat

Experiences over the past 50 years in a variety of conflicts have demonstrated that unmanned aerial systems, including cruise missiles and unmanned aerial vehicles (UAVs), can play versatile and effective roles in offensive military operations. The development of the modern, highly accurate land attack cruise missile in the 1970s and 1980s, along with the use of such missiles in the First Gulf War, has cemented the place of cruise missiles in the thinking of offensive military planners. Likewise, the use of UAVs in the recent wars in Afghanistan and Iraq has made UAVs a prominent element in the thinking of contemporary military planners as well.

UAVs and cruise missiles represent important technological tools in the U.S. arsenal, and they fulfill a variety of functions from persistent surveillance to precision attack. Their utility for military applications is not new; however, changes in the international environment stemming from the collapse of the Soviet Union, the rise of a virulent form of terrorist organizations willing and able to strike the U.S. homeland, and the increasing availability of critical technologies for such systems—and of the systems themselves—have raised questions about whether the relative threat posed to the United States by adversary use of such systems is changing enough to warrant greater attention.

The availability of UAVs and cruise missiles to potential attackers will be determined to a large degree by their availability to legitimate military and civilian users around the world. The most attractive feature of UAVs and cruise missiles for a nation-state is that a viable offensive air capability could be developed at a small fraction of the cost of more-conventional systems, such as piloted fighters. Furthermore, cruise missiles require far fewer day-to-day maintenance expenditures, and a less well-trained force can operate them effectively. In this regard, it appears that UAVs and cruise missiles are likely to become relatively ubiquitous as military systems.

UAV systems are becoming increasingly capable and available with time. Some types of UAVs are available for purchase "off the shelf," and systems with limited capabilities, such as technologies in hobbyist-driven markets for remote-control vehicles, have long been widely available at low cost. Now, however, systems with larger payload capacities and capabilities are becoming more readily available. Once restricted to the military arena, UAVs are finding commercial and civil applications. The civilian UAV market is still in its infancy, but the smaller UAVs already are emerging in a unique niche, providing commercial imagery from the air, in the scientific arena, and even in law enforcement.

A report available from the U.S. Naval Institute's Periscope Web site suggests that, over the next ten to 15 years, there will be a demand for 3,000 moderately stealthy long-endurance UAVs, 2,000 Eagle II–class and Predator A/B UAVs, approximately 500 8,000–15,000-lb UAVs, and hundreds of thousands of micro-UAVs ("The High-Flying UAV Marketplace," 2004; Goshen-Meskin, 2005). A 2007 assessment in *Aviation Week & Space Technology* suggests that the market will be worth $16 billion over the next ten years. Of this market, more then 60 percent will be in the hands of U.S. companies; European companies will hold about 6.5 percent; Israeli companies, 2.6 percent; and the rest of world (including Russia), about 4.7 percent (Dickerson, 2007). Given the low cost of market entry, many other countries have small and medium-sized UAV programs. The same report indicates that, although the United States accounts for 75 percent of research and development, many other countries are entering into the UAV business. Countries as diverse as China, India, Pakistan, Iran, Japan, Syria, and Australia are developing and/or operating UAVs, all of which can be expected to sell into the international marketplace (Zaloga, no date). The total number of countries believed to be developing some type of UAV is 18, 13 of which are currently exporting the systems (Bolkcom,

2002, p. 15). In addition, 22 other countries are reported to be capable of transitioning to producing cruise missiles.

As with the UAVs themselves, the costs of these systems vary broadly. Those of Western military UAVs range from a few thousand dollars for small UAVs to more than $67 million for the largest UAVs, such as Global Hawk, with ground equipment factored into the cost (U.S. Government Accountability Office, 2005). Systems designed for civilian applications are cheaper, but they also vary from application to application. The costs of these systems are typically in the low thousands to few tens of thousands of dollars, and the air vehicle is only a fraction of the total cost. For example, costs for the Japanese RMAX rotary-wing UAV range from a low of $86,000 to a high of $1 million for a fully autonomous system with two airframes, a base station, and camera systems.[1] The Aerosonde UAV costs are reported to be around $25,000 (McGeer and Vagners, 2000). Small UAVs designed for police applications cost upwards of $8,000, with ground stations adding another few thousand to the price.[2]

Homegrown UAVs originating from the model-aircraft market represent another possible source for an adversary seeking a basic UAV capability. Most model aircraft of interest would cost several hundred dollars. Such a kit would be capable of line-of-sight operation; one with full autonomy through the addition of autopilot systems could be expected to cost a few thousand dollars. The larger the basic airframe is, the greater the expected costs of the aircraft would be.

As these markets develop, UAV systems will be produced in larger numbers, at lower cost, and with a wider variety of capabilities than are available today. These changes, occurring to a great extent independently of military technology applications, will increase these systems'

---

1 The RMAX is one of a family of rotary-wing UAVs used for crop dusting and aerial photography in Japan ("Yamaha's RMAX—The Worlds Most Advanced Non-Military UAV," no date).

2 An example is FARSIGHT Intelligence Systems' RAIDER, which is priced at just less than $8,000 (see the FARSIGHT Products homepage), and the SkySeer UAV, which costs a reported $25,000 (see Bowes, 2006). Note that the experiment using the SkySeer encountered problems with the Federal Aviation Administration (FAA) for operating within the air traffic system.

availability to a range of state and nonstate actors. Although many current UAVs focus on intelligence, surveillance, and reconnaissance (ISR) applications, the vehicles can also be used to deliver attack payloads to a target.

For cruise missiles, the proliferation of a number of key technologies on which the weapons depend for their capability—integrated satellite/Inertial Navigation Systems (INS)—has significantly reduced the obstacles for developing, fielding, planning, and effectively employing such weapons.[3] The ability to convert existing missile systems designed for other purposes—such as anti-ship cruise missiles—into land attack cruise missiles provides an additional route for an adversary to develop a limited cruise missile capability. The Missile Technology Control Regime (MTCR) puts some limits on nation-states' acquisition of these weapons; however, a variety of approaches have been adopted by nation-states to limit the effect of the restrictions. For example, one strategy has been to develop missile systems that, when manufactured, are technically in compliance with MTCR restrictions on weapon range but that could have their range readily extended should the producing country wish to do so at a later date. Over time, these shifts will continue to increase the variety of cruise-missile technologies available and the range of potential applications of these weapon systems.

Cruise missiles have a somewhat smaller set of suppliers than UAVs. Russia, China, and the United States produce most such missiles. The total worldwide inventory of cruise missiles is on the order of 80,000, 14,500 of which were reported to have been exported from the producer nations (Systems Assessment Group, NDIA Strike, Land Attack and Air Defense Committee, 1999). The vast majority of the missiles are anti-ship missiles, but a moderately sophisticated opponent might convert some of them to land-attack mode. In recent years, small numbers of anti-ship cruise missiles apparently have been directly transferred to quasi-state subnational groups (e.g., transfer of the

---

[3] The best known of these are the Global Positioning System (GPS)/INS combinations that the United States uses. Certainly, similar integrated satellite/inertial navigation systems are possible using other satellite systems, such as the European Galileo, Russian GLONASS, or Chinese Beidou satellite navigation systems.

C-802 anti-ship missile from Iran to Hezbollah), and have been employed in combat (see Myre, 2006). Western cruise missiles are available for on the order of $0.5 million at the low end to more than $2 million on the high end for the most-capable systems.[4] Non-Western systems are generally believed to be offered at lower prices. It is reasonable to expect that the floor for new long-range missiles would be a few hundred thousand dollars. It is not clear what the costs are of converting an anti-ship missile for land attack or what the fixed costs of those efforts would be to a country converting the missiles.

The expanding availability of UAVs and cruise missiles,[5] coupled with their increasing capability and versatility, has led to concern about how potential adversaries might use these technologies, particularly as part of asymmetric[6] strategies. As attack platforms,[7] UAVs and cruise

---

4 The conventionally armed Tri-Service Standoff Attack Missile cost had risen to more than $2.1 million (1994 dollars) (U.S. Government Accountability Office, 1996) prior to cancellation, and the nuclear-tipped Advanced Cruise Missile cost was estimated at more than $2 million (1993 dollars) (Forecast International, 2003).

5 To represent a threat, a UAV or cruise missile that is acquired internationally and is intended for use in the United States would have to be brought into the country. Although there is significant concern about the ability of individuals or groups to bring a variety of materials into the country illegally (e.g., drugs, smuggled goods, individuals), there is some risk of apprehension at the border. For example, in Sri Lanka, two UAVs were seized at the border (Warnakulasuriya, 2003).

6 The concept of *asymmetry* has been used in a wide variety of ways in discussing threats and security situations and, as a result, has a variety of meanings and connotations in different parts of the literature (see Lambakis, Kiras, and Kolet, 2002, for a review of the use of the term across a range of contexts). In the current work, we are using the term *asymmetric* to connote use of UAVs and cruise missiles in operational applications that differ from their use in standard military-on-military engagements during overt hostilities. This use includes that by nonstate groups for strikes on civilian and other targets away from defined theaters of hostilities (most specifically, for strikes on targets in the U.S. homeland), as well as by state actors either to attack non–front-line military targets during a conflict with the United States (e.g., strikes on out-of-theater U.S. military installations or attacks on U.S. allied states during coalition warfare operations) or to carry out strikes on targets in the homeland for terror or other influence purposes.

7 UAVs in particular can be applied to a variety of activities beyond use in direct-attack scenarios: mainly, use as intelligence, surveillance, and reconnaissance platforms. Our study focused on direct-attack scenarios for both UAVs and cruise missiles and did not examine other applications for these platforms.

missiles can be applied to a wide range of operational scenarios, and the nature of U.S. society and infrastructures necessarily means that a wide variety and large number of targets in the homeland are vulnerable to aerial-attack scenarios that could be facilitated by UAVs or cruise missiles.

The published literature discusses scenarios of concern that include attacks on infrastructure targets, such as water or power plants, on dense crowds, or for the dispersal of unconventional weapons (Verton, 2005, pp. 10–14; Gormley, 2003, pp. 3–9; and Miasnikov, 2005). The apparent plausibility of such attack scenarios—particularly in the current environment of heightened concern about attacks by nonstate groups or state adversaries adopting asymmetric strategies against the United States—has led a variety of analysts to examine defensive options to protect the United States from attacks using UAVs and cruise missiles. This work has highlighted problems with detecting these types of small, low-flying targets; how to appropriately respond to a threat detected over a populated area; and the resources and capabilities needed to defend the airspace of a nation as large as the United States (Gormley, 2006; Gormley, 2003, pp. 3–9; Bolkcom, 2006; and Miasnikov, 2005).

The RAND Approach

To understand how UAVs and cruise missiles contribute to the spectrum of threats faced by the U.S. homeland, RAND conducted a study assessing the potential use of these platforms in asymmetric attack scenarios in the United States. Building on previous analyses of these systems and a body of RAND work[8] on state and nonstate group behavior and technological decisionmaking, these systems were assessed from the perspective of actors planning asymmetric operations. The study approach was based on two fundamental principles: (1) The threat posed by UAVs and cruise missiles cannot be assessed in isolation and

---

[8] Recent examples include Jackson, Chalk, Cragin, Newsome, Parachini, Rosenau, Simpson, Sisson, and Temple (2007); Daly, Parachini, and Rosenau (2005); Cragin and Gerwehr (2005); Cragin and Daly (2004); Jackson (2005a); Jackson, Baker, Cragin, Parachini, Trujillo, and Chalk (2005); and Jackson (2001, pp. 183–213).

(2) defensive options to address the threat must be considered broadly. These principles are reflected in two study tasks.

Task 1: Conduct a Red Analysis of Alternatives

In pursuing their goals, terrorist organizations and other potential adversaries frequently consider a range of options. Therefore, the attractiveness of these technologies will be driven not only by their characteristics but also by their relative advantage over other attack options, leading us to a "red analysis of alternatives," in which we weigh UAVs and cruise missiles against alternative ways in which potential terrorist groups might attack targets of concern.

Whereas an *analysis of alternatives* considers the benefits, costs, and risks of different options in order to select the best option, a *red analysis of alternatives* performs a similar comparison of options from the point of view of a potential adversary. For several types of attacks, we compared the suitability of cruise missiles and UAVs against other options, such as vest bombs, car bombs, and mortars. In doing so, we do not imply that a terrorist group will necessarily undertake a rigorous or quantitative assessment of the separable costs, benefits, and risks of different attack options in operational planning but, rather, that decisionmaking will focus on choosing attractive attack modes from the terrorists' point of view and criteria. Such an assessment could be intuitive or implicit rather than quantitative and methodical.

The benefit of such an approach does not depend on replicating the decision process of a particular adversary; instead, by identifying the operational problems faced by a potential adversary, it helps the defense understand how the capabilities of different attack modes could help an attacker overcome those problems. It is also not "mirror imaging," or mistakenly assuming that your adversary will make identical decisions you would make in a given situation, because we are not seeking to predict exactly how an adversary will act.

Task 2: Assess the Implications for the Defense

Just as Task 1 considers a range of attack options, the assessment of defensive options must be broad as well. In crafting defensive solutions for specialized threats, it is important to maintain a broad view of defensive options and how individual defenses perform in the context of an overall spectrum of threats.

In seeking concepts to guide prudent resource allocation for defensive measures, we considered defensive options across the full range of adversary activities, including activities before, during, and after an attack, rather than a preferential focus on classical "terminal defense" strategies. Recognizing the many demands on the resource options for homeland security, we sought defenses that provide common protection against both this and other asymmetric threats within reasonable cost constraints.

About This Report

Chapter Two of this document examines UAVs and cruise missiles and compares them with other available attack modes. Chapter Three assesses five key operational problems for which UAVs and cruise missiles appear to be desirable solutions and assesses available alternative ways in which attackers might solve those problems. Chapter Four discusses adversary preferences and organizational characteristics that could shape the attractiveness of UAVs and cruise missiles as chosen weapons. Chapter Five discusses strategies for defending against these threats. Chapter Six concludes with a discussion of cross-cutting lessons about these threats and the assessment of novel threats in general.

CHAPTER TWO

UAVs and Cruise Missiles as Asymmetric Threats: How Do These Systems Compare with Alternative Attack Modes?

The cruise-missile and UAV industries are very dynamic. New systems and new applications are designed each year for commercial and military applications. Terrorist groups have not ignored these systems; however, that only a few examples of terrorist experimentation and use have come to light to date suggests that interest is not widespread. In this chapter, we outline the UAV and cruise-missile markets and their distinguishing characteristics and describe the use of these weapons in three broad attack modes. Finally, we compare these three modes generally with other attack modes to highlight their distinguishing attack characteristics.

The demonstrated utility of cruise-missile systems for military applications has generated significant interest from a variety of nation-states, including China, Russia, France, India, and Iran. Many nations are developing tactical UAVs capable of supporting battlefield intelligence needs; a smaller set of nations is developing UAVs suitable for strategic intelligence and warning functions in their theater of operations. Very large, high-endurance UAVs, such as Global Hawk, are being pursued in a few countries that have significant interests over large areas of the globe and wish to support intelligence and military missions far from their national borders.

Some nonstate groups have also shown interest in unmanned aerial vehicles to enable attack: Hezbollah has demonstrated UAV capability on two occasions with overflights of Israel (see, for example, Gormley, no date; and Karmon, no date); Hamas attempted to procure UAV technology for offensive attack applications (Jane's Terrorism and Insurgency Centre, 2003); the Revolutionary Armed Forces of Colombia (FARC) retrofitted (though never used) model airplanes with explosive payloads ("Troops Seize Rebels' Explosive Planes," 2002); and discussion of the use of unmanned aerial attack modes has occurred in other terrorist Internet forums ("Al-Qaeda Online: Understanding Jihadist Internet Infrastructure," 2006).

Other analysts have cited additional instances of either expressions of interest or pursuit of these weapons by such groups (see, for example, the discussion in Gips, 2002). The 2006 conflict between Hezbollah and Israel demonstrated broad use of some types of rockets and ballistic missiles by a nonstate organization, including weapons with ranges reaching to and potentially beyond 100 kilometers (km) (see, for example, the discussion in Gardner, 2006). Hezbollah also used a limited number of anti-ship cruise missiles against naval targets, including a successful strike on an Israeli military vessel (see the discussion in Hilburn, 2006). Cruise missiles, if they became readily available[1] to these organizations, could be expected to attract some interest, provided they could be operated at acceptable levels of operational risk and transported within range of their targets. The interest in aerial attack systems by an increasing number of actors with disparate motivations has broadened the variety of potential threats to U.S. interests posed by such systems.

Whether UAVs and cruise missiles will be attractive to a particular adversary will be driven in large part by the group's goals, whether these systems are compatible with those goals, and what advantages these systems deliver over other ways of doing the same things, many of which will necessarily be more familiar, cheaper, and more certain technologies than the use of UAVs or cruise missiles.

---

[1] The access of nonstate actors to UAV and cruise-missile technologies could clearly be shaped by their relationships to states. States could assume the role of technology provider (by state policy, through action by specific government organizations or individuals with access to the technologies, or by theft of the technology from state arsenals) to such organizations. This is believed to be the case for Hezbollah: Iran is thought to have provided the group with a number of UAVs and training in their use. See Verton (2005, pp. 10–14). This issue is discussed in Chapter Four.

The goals that stand behind the violent actions of individual adversaries can obviously differ considerably. For example, even if they staged very similar operations, a hostile state and a terrorist group could be pursuing very different fundamental goals. The goals of offensive action can relate to the specific target of the violence or to the reaction to the strike in other populations or audiences.

Whereas an actor may have a specific goal that it is trying to accomplish through violence, there may not be a simple and direct relationship between that goal and the violent operations it conducts, whether utilizing a UAV, cruise missile, or any other mode. Actions for which the goal of the attack is very instrumental—e.g., an actor is concerned with the United States' moving troops in an area via a specific troop transport—the relationship can be clear and direct. An attack that disables or destroys the transport will achieve the goal of the adversary. In other cases—and in the majority of cases relevant in a consideration of asymmetric warfare and terrorism—the relationship between the violence and the goals an adversary is trying to achieve may not be direct. Violence produces specific "tactical outputs":

- Targeted individuals are injured or killed.
- Property is damaged or destroyed.
- An activity in or by the targeted state is disrupted.

Where the outputs themselves do not directly achieve the adversary's goals, they must be somehow linked to achieving those goals. Will killing specific individuals or large enough numbers of people result in changes in the targeted state's behavior? Will economic costs from property damage or disruption hurt its stature nationally? If the goal is to impress or appeal to a specific audience, what is needed to do so? Will simply demonstrating a new and powerful attack be sufficient, or will it be necessary to produce specific types or levels of damage?

Subsequent decisions made by the attacker about targets and attack modes will be driven significantly by what tactical outputs it believes are useful for its purposes. Choice of location will also be part of the calculus: Even if an adversary's ultimate target is the United

States, it might choose to stage operations against interests outside the United States or against key U.S. allies or partners.

Characteristics of UAVs and Cruise Missiles

The need to meet different operational demands has led UAVs and cruise missiles to be developed with differing characteristics, such as range, flight speed, payload capacity, and other capabilities. Differences in the characteristics of available systems arise from the different missions they have been designed to carry out:

- *Cruise missiles* are exclusively attack platforms and have been designed for rapid penetration of defensive measures and delivery of a large, high-explosive warhead, and for launch from a variety of platforms in a range of military scenarios. These systems are therefore generally characterized as having large-payload capacities, fast flight speeds, and ranges that vary from short (tens of nautical miles [nmi]) to long (more than 1,000 nmi).
- Most currently available *UAVs* (with the exception of systems that carry missiles to strike targets) have not been designed as direct-attack platforms. They have generally been optimized for ISR applications and have been designed around payloads of sensors and communications equipment. Many larger UAV systems have been shaped by the desire to have them stand off and loiter near targets for extended periods to provide surveillance coverage; therefore, they have been built with extended flight times (and, therefore, ranges) in mind. UAV systems at the other extreme of the size scale (e.g., systems designed to provide military forces with tactical, "over-the-next-hill" visibility and reconnaissance capabilities) have been designed for easy deployment but with very limited ranges and flight times. After examining the full spectrum of UAV systems, we can report that ranges vary from 1 nmi or less to thousands of nautical miles for the longest-range systems. Because of their design around sensor and communications packages, many of the UAV systems have more-limited pay-

load capacities (e.g., some are able to carry less than a pound [lb] of added weight) than cruise missiles.

As UAVs adopt more attack missions, they will increasingly begin to share the operational characteristics of cruise missiles. But, for now, clear distinctions can be seen in comparison. Figure 2.1 summarizes relevant characteristics of currently available UAVs and cruise missiles: useful payload, range, system weight, and speed. Note how distinct the cruise missiles (depicted as blue squares) are from the population of UAVs (shown as red dots). Note also that, while the payloads of many types of UAVs overlap those of cruise missiles, the total numbers of UAVs tend to be dominated by smaller UAVs designed for tactical applications and produced in much larger numbers. For example, the United States Marine Corps plans to procure around 1,400 small

Figure 2.1
Speed, Platform Weight, Range, and Payload Capacity of Currently Available UAV and Cruise-Missile Systems

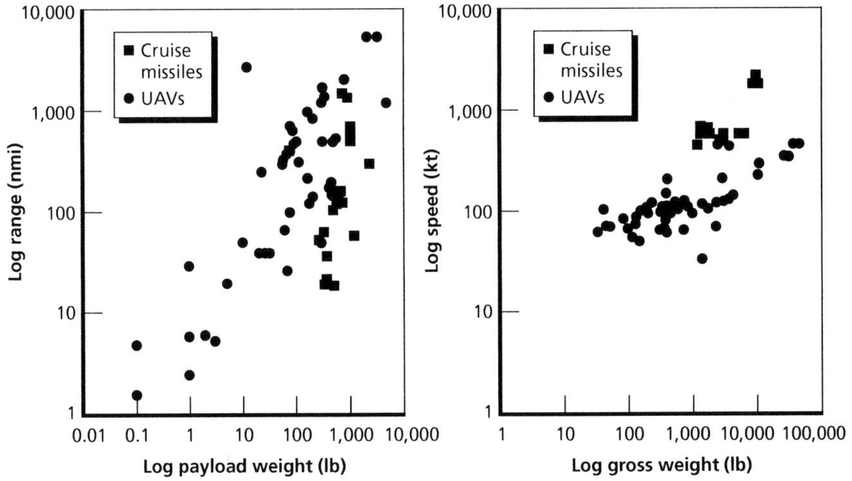

SOURCES: U.S. Naval Institute, Periscope database; U.S. Department of Defense, Office of the Secretary of Defense, *Unmanned Aerial Systems Roadmap 2005–2030*, Washington, D.C., 2005.

RAND *MG626-2.1*

Dragon Eye aircraft, whereas the United States Air Force is looking at procuring 51 of the far larger Global Hawks (U.S. Department of Defense, 2005).

Comparing the Capabilities of UAVs and Cruise Missiles in Attack Scenarios to Those of Alternative Attack Modes

UAVs and cruise missiles could be used in attack scenarios in a wide variety of ways, which can be divided into three broad classes (Figure 2.2):

- *Direct*, wherein the UAV or cruise missile is employed by striking the target with the weapon and damaging it through the force of impact and the effect of any payload carried by the vehicle. This is cruise missiles' primary attack mode. Direct-attack scenarios are relevant for a variety of targets, from fixed sites (e.g., infrastructure or buildings) to mobile targets (e.g., vehicles, crowds, or individuals.) Unconventional weapons, including chemical, biological, and radiological (CBR) agents or nuclear devices, could be used as payloads in direct-attack modes.
- *Indirect or "challenge response,"* in which the aerial system is used to produce a reaction on the ground without directly attacking the target (e.g., individuals evacuating a building in response to fear that the UAV will strike the structure). In this case, the operational goal could simply be disruption (the response itself is the desired outcome); or the goal could be to use the response to enable a follow-on component of the operation (e.g., attacking evacuating crowds with bombs prepositioned in the evacuation zone).
- *Aerial dispersal*, in which the UAV releases a payload (either nondestructively or destructively) over one or more targets below it. This class is most relevant for scenarios involving chemical, biological, and radiological or nuclear weapons, where release at altitude could increase the effectiveness of an attack.

Figure 2.2
Three Classes of Offensive Application Modes for Unmanned Aerial Vehicles and Cruise Missiles

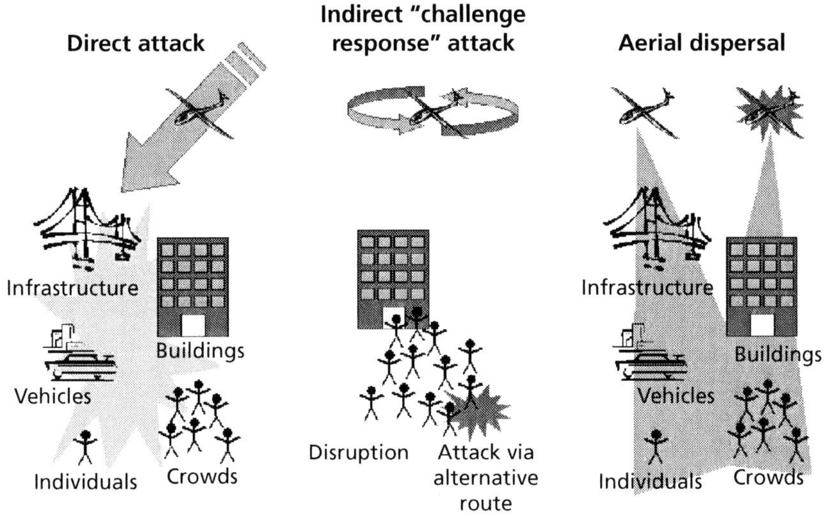

RAND *MG626-2.2*

However, for any potential target, there are a variety of ways an adversary might stage an attack, and UAVs and cruise missiles represent only a subset of the possible *aerial* attack modes. A variety of other options are available to attackers:

- *Direct-attack* alternatives include ground-based attacks, such as emplaced bombs, suicide bombs, snipers, armed assaults, or (for unconventional weapons) ground-based or manual dispersal modes; aerial scenarios, such as suicide or nonsuicide uses of aircraft as weapons; or the use of indirect-fire weapons, such as mortars or rockets, to deliver either conventional or unconventional payloads.
- *Indirect-attack* modes, intended to produce reactions at a targeted site, include information operations, such as bomb or other threats; creation of an apparent or real emergency (e.g., triggering a fire alarm, planting a small bomb at the scene aimed at generating a reaction rather than producing casualties); or taking action

to simulate an attack in progress or threatening behavior (e.g., penetrating restricted airspace with an aerial vehicle).
- *Aerial dispersal* of weapon agents include use of airbursting weapons, such as mortars or rockets; use of traditional aerial vehicles, such as planes, helicopters, or balloons; or dispersal from an elevated location (e.g., geographic feature, tall building).

The choice of a UAV, cruise missile, or an alternative attack mode will depend on how well the characteristics of the tactic or system match the adversary's goals for the attack.

Assessment of the relative attractiveness of UAVs and cruise missiles to asymmetric adversaries requires more than simply listing potential alternative attack modes that could be used instead. Real comparisons among alternatives require identifying the primary characteristics of these systems and of alternative approaches that dictate their effectiveness. For an individual state or nonstate actor, the attractiveness of specific systems is determined by factors beyond just their relative effectiveness, which essentially addresses only the benefit part of a benefit/cost/risk judgment.[2] System characteristics that affect perceived costs and risks are more idiosyncratic to particular groups and are discussed in Chapter Four on group-specific characteristics affecting such analyses. The following subsections compare the technical characteristics of different attack modes for purposes of conducting direct-attack, indirect-attack, and aerial-dispersal operations.

Direct Attack

In direct-attack scenarios, the primary determinants of success are the warhead effectiveness (approximated by the weight of the payload delivered to a target), the type of ordnance (i.e., blast, fragmentation

---

[2] Note that the assessment of the benefits, costs, and risks of an adversary adopting a particular weapon may be an implicit rather than explicit process. The process will invariably be complicated by uncertainty and imperfect information. Different adversaries will also have preferences and biases that will shape some conclusions about weapons and targets. Some planners might approach the planning process as nearly fully rational actors; others might have strong preferences that distort how they perceive the benefits or costs of courses of action.

weapons, or an unconventional payload) that is delivered, the accuracy of the weapon, and the probability of arriving at the target.

A static comparison of payload capacities indicates that UAVs and cruise missiles are quite similar in payload to other systems. The smallest UAVs have practical payloads in the subkilogram (kg) range, and the largest are in the 500–1,000-kg range. In comparison, small backpack and vest bombs have charges in the 5–15-kg range, vehicle bombs are on the order of 500 kg, and large truck bombs are in the 1,000–15,000-kg range. Aerial methods of attack have small mortar bombs in the 1–2-kg range, and conventional aircraft have maximum useful payloads of a few hundred kilograms for light aircraft, a 1,000-kg useful payload for business jets (bizjets) (beyond the aircraft itself), and on the order of 100,000 kg of payload for large transports.[3]

Different targets are vulnerable to different types of damaging payloads. For instance, people are vulnerable to both blast and fragmentation, whereas heavy structures are generally quite resistant to the impact of fragments. UAVs, cruise missiles, and most alternative modes can deliver payloads causing both blast and fragmentation damage, although incorporation of material and shrapnel for fragmentation reduces the total weight of explosives that can be used, limiting blast damage. Small-payload weapons, such as mortars, rely primarily on fragmentation for their destructive effect. Unconventional payloads (with the exception of nuclear devices) in direct-attack scenarios are most relevant for striking at people, although they can produce contamination at targets as well. In such scenarios, success is driven by how well the agent is dispersed and the presence of vulnerable human targets in the dispersal area.

For purposes of illustration, we consider an attack that might be attractive to a variety of asymmetric adversaries: a strike against a population gathering, with the goal of inflicting serious injury or fatalities on the group. For analytic purposes, we use a 10-pounds-per-

---

[3] Weaponization of large aircraft can be difficult because of the difficulty of triggering such large charges. As with the attacks on September 11, 2001, many attacks might just depend on the kinetic effects of the aircraft and damage created by subsequent fires from residual fuel.

square-inch (psi) threshold for blast effects capable of creating serious injury to illustrate the impact of varying unitary-charge sizes on an attack against a crowd located uniformly in a 10,000-meters-squared (m^2) open area.[4] For comparison with the unitary charges, we use the fragmentation area for a typical modern mortar round. The bursting radius, which illustrates the high-probability area for wounding by the mortar round, is based on a typical medium-weight mortar round. The charge sizes selected for comparison are described in Table 2.1.

Table 2.1
Estimated Representative Explosive-Charge Sizes for Relevant Weapon Types

Weapon	Warhead Weight (Kilograms of C-4 Equivalent)
Bomber vest	9
Car bomb (low)	100
Car bomb (average)	225
Car bomb (large)	500
Truck bomb (light)	1,500
Truck bomb (medium)	3,000
Truck bomb (large)	15,000
Small UAV	5
Medium UAV	100
Large UAV	500
Bizjet	900
Cruise missile	500
Mortar	

SOURCES: RAND estimates derived from literature sources (e.g., Bureau of Alcohol, Tobacco, and Firearms, "ATF Vehicle Bomb Explosion Hazard and Evacuation Distance Tables," Washington, D.C.: ATF Instruction 5400.1, January 1999; and expert input.

---

4 For a more detailed explanation of warhead effects, see "Warheads" (1998), and Glasstone and Dolan (1977, pp. 548–559). We used a 10-psi threshold, well within typical fragment-injury areas of a conventional explosive, to capture the low end of blast effects.

The different potential attack modes for an attack on a gathered population are presented in Figure 2.3 for comparison. The effective radius of the weapons is indicated by blue bars plotted on the *x*-axis to the left of the figure, and the corresponding fraction of the total 10,000-m^2 target area is indicated by red circles and plotted on the axis to the right.[5] The most interesting thing that can be seen from the cases shown is that the UAV-based options with high-explosive warheads

Figure 2.3
Comparison of Alternative Attack Modes

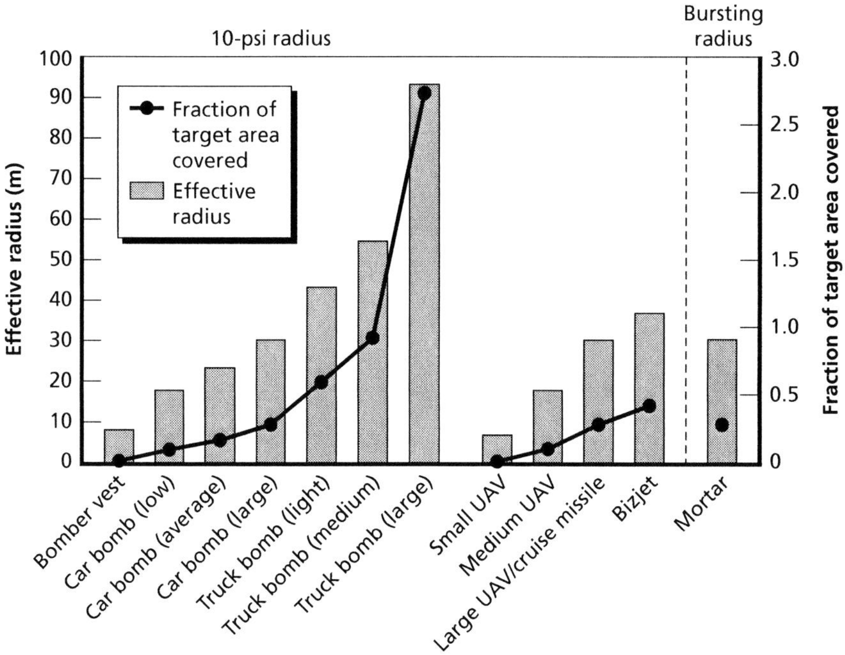

SOURCE: RAND analysis.
NOTE: Overpressures estimates assume unitary explosives.
RAND *MG626-2.3*

5 Based on the characteristics of the weapons, coverage of the target area ranges from less than a tenth of the area for a suicide bomb vest to more than two-and-a-half-fold coverage with a large truck bomb.

resemble the smaller terrestrial-attack options in terms of area coverage, but they fall well short of the effects of the largest weapons that are available for attacks on the ground. Also, if the attacker considers wounding by fragments a satisfactory outcome (which could be produced, for example, by a mortar that can be airburst over the crowd), then small weapons such as mortars (or small UAVs with similar warheads) could cover a significant fraction of the area.

Many more variables affect the use of unconventional weapons in such a scenario, including the point of release, local atmospheric and other conditions, and the nature of the weapon. Such weapons could potentially cover a much larger percentage of the target area, although the outcome of doing so (e.g., casualties produced, other costs of the attack) is more difficult to anticipate than for conventional payloads.

Among these varied modes of attack, there are also important differences in the probability of reaching the target in the first place and the potential accuracy of that delivery. In most cases, ground-based attack modes employing vehicles have lower probabilities of getting to the target,[6] but they will have greater accuracy and higher maximum payloads than nonterrestrial options. Ground-based systems employing pedestrian suicide bombers enjoy a higher probability of reaching many targets, but they have a far smaller payload. Attacks that seek to go over defenses, depending on the weight of the payload of the system, may have a somewhat lower per-weapon effectiveness[7] than ground-based systems but a much higher probability of getting into the general target area. And such systems as mortars have a multivolley, multidirectional capability. Similar arguments can be made for direct-attack scenarios involving unconventional weapons.[8]

---

6 Lower probabilities result because these attack modes would be subject to ground-based security measures that could prevent access to a desired target.

7 Smaller payloads in such systems may require trading off wounding potential from fragmentation against killing potential from blast effects from the explosive charge itself.

8 The attractiveness of aerial-delivery modes for direct attack using unconventional weapons will be driven significantly by the nature of the target being struck. For example, if the desired target is the population inside a structure, the potentially higher probability that the UAV or cruise missile will reach the target (e.g., by avoiding security around the building) might be offset by the chance that the weapon will not actually penetrate the structure before

How trade-offs are made among these modes/weapons will depend to a great extent on the preferences of the attacker: Some attackers may optimize as they choose among alternatives, attempting to produce the most damage they can for a given attack or the most certainty of attack success. Others may satisfice, seeking a "good enough" outcome for their attack, to meet whatever thresholds they believe are relevant for accomplishing their goals.

Indirect Attack

The effectiveness of indirect-attack modes relies on the ability of an attacker to produce predictable reactions so that a target moves into a vulnerable position or to generate a desired level of disruption. While successfully anticipating such behavior is difficult—a challenge for any attacker—the effectiveness among modes will be determined by how consistently the modes produce the actions desired by the adversary.

It is difficult to generalize about the relative effectiveness of indirect-attack modes, and the modes' effectiveness may differ from target to target. For example, phone calls or other threats have historically proven useful to trigger evacuation of people from the immediate area of a target; once they are evacuated, the people become more vulnerable to attacks on the street than when they were in the building. Telephone calls have similarly been used to attract emergency responders to a scene as potential targets of attack. The past use of this type of provocative attack has forced many authorities to be wary of bomb warnings and to weigh the risks of staying in a potential target against the risk of driving people into a potentially more dangerous environment outside a security perimeter. The effectiveness of such strategies will depend on the security posture of the target, however. For example, such strategies might work far less effectively at defended government targets than at accessible public sites.

Indirect attacks using UAVs are a possibility for an attacker. The use of UAVs in this role has not been demonstrated, but on the face of it, it would seem possible if the potential targets have previously dem-

---

releasing its payload. Release of a weapon on the outside of a targeted structure is unlikely to produce the effects that an attacker desires.

onstrated that they trigger an evacuation in response to potential air attack. For example, the threat of an air attack drove the evacuation of the U.S. Capitol complex before the state funeral for Ronald Reagan ("Capitol Evacuated Before Reagan Procession," 2004). UAVs or other types of aircraft could be used to drive people out of cover, exposing them to attack from other means, such as prepositioned bombs or gunfire. Whether or not such an indirect challenge would be effective would ultimately depend on the defender's actions and consequently, would have a degree of uncertainty. Consistent behavior on the part of the defender (say, always evacuating based on any type of air attack, regardless of the observed attack system) would decrease the attacker's uncertainty and might make this approach fairly attractive.

A group relying on indirect strategies as a long-term tactic is questionable, given that the targeted individuals may take the opportunity to make changes in their behavior. Such an approach is most likely to be effective when the threats used to produce the response are highly credible (e.g., use of bomb threats during an ongoing bombing campaign, which makes it difficult or impossible for the defender to distinguish a threat from an actual attack) or when the mode used to produce the indirect response is unexpected. The main potential advantage of UAVs or cruise missiles would be their unexpectedness: It is one thing to receive a telephone bomb threat from an unknown caller; it is something else to have a message coming through official channels indicating that the building you are responsible for may be subject to air attack.

The comparison of these systems in direct-attack scenarios discussed above suggests that their relative effectiveness is comparable to that of other attack modes; however, the novelty and psychological effect of air attacks may increase their utility in indirect-attack scenarios. However, unless a group carried out many effective attacks using these systems, thereby maintaining the credibility of the potential threat they posed, their utility in indirect strategies would likely decay over time.

Aerial Dispersal

The effectiveness of modes for dispersing an attack agent in the air above a target relies on the ability to place sufficient amounts of the weapon in the desired position, its probability of arriving there successfully and at the time designated for the attack, and the chance of successfully dispersing the material in the manner desired. Aerial-dispersal modes for such payloads as biological agents or radiological material have been viewed as one mission that is particularly attractive for UAVs and cruise missiles. Particularly for UAVs, the systems' ready availability and ability to fly in most areas that would represent attractive targets appear to be significant advantages.

However, they have significant disadvantages in payload size and the probability of successfully deploying the agent at the position and time desired. While UAVs and cruise missiles can certainly carry operationally relevant unconventional payloads (particularly biological and radiological weapons), they do so at greater operational risk than many more mundane delivery mechanisms. They are subject to significant uncertainties from the effects of weather on their remotely controlled flight than are piloted aircraft. As a result, a general aviation aircraft has a significant advantage over most UAVs in reliability and payload if a pilot is available. Similarly, UAVs and cruise missiles must rely on technological mechanisms to disperse the agent at altitude; such mechanisms could fail, especially during complex dispersal operations (e.g., extended line release instead of a simple point release). Although the use of piloted systems would risk exposing the flight crew to the agent, such exposure would not deter some terrorist organizations.

Conclusions

As attack systems, UAVs and cruise missiles deliver to targets payloads that can range from very small (equivalent to the explosives content of a mortar round, for example) to relatively large (on the order of the explosives contained in a car bomb). These payload capacities are comparable to or below those of other ways in which state or non-state groups can deliver weapons to targets. The same can be said for indirect-attack and aerial-dispersal applications, in which less-sophisticated systems and tactics deliver comparable or potentially superior capability.

The advantage that UAVs and cruise missiles provide is not, therefore, in the destructive power that they can carry; rather, it is in the way they carry it and the distance from which they allow an adversary to control its delivery. The value of this advantage to an adversary and, as a result, the likely attractiveness of these systems will therefore be driven by the benefits of aerial attack in solving specific operational problems. To explore situations in which UAVs and cruise missiles might be most attractive to asymmetric attackers, we examined a set of operational problems for which aerial capability appeared potentially most valuable and compared these systems to other alternative solutions.

Whereas cruise missiles and UAVs have different characteristics, which have been compared in this chapter, for the remainder of the report we shall consider them as one class of threat without distinction.

CHAPTER THREE

What Adversary Operational Problems Can UAVs and Cruise Missiles Best Solve and How Do UAVs and Cruise Missiles Compare with Alternative Solutions?

When the *effects* that can be produced by payloads delivered by UAVs or cruise missiles are examined against those produced by the range of alternative attack modes available to adversaries, UAVs and cruise missiles do not appear to stand out as meriting particular attention for use against undefended targets. As long as options such as suicide operatives or vehicle bombs can be used, these more-basic and more-reliable means will generally make it possible to deliver more potent payloads to desirable targets.

Their destructive similarities notwithstanding, the fact that UAVs and cruise missiles enable *aerial* attack does make them stand out, particularly for adversaries that might not otherwise have the ability to attack from the air. The ease of launch and potentially long-duration flight for some of these systems can be a major capability improvement for some adversaries, particularly nonstate groups. As a result, this analysis focuses on the most promising cases in which a cruise missile or UAV might provide an adversary with a sufficiently large advantage that these particular delivery modes might merit disproportionate concern.

UAVs and cruise missiles are most likely to be attractive in situations in which their aerial, long-standoff capability solves key operational problems an attacker faces in planning and mounting an operation. These systems appear most advantageous because they enable solutions to five main problems (summarized below and in Figure 3.1).

Figure 3.1
Asymmetric Capabilities Enabled by Unmanned Aerial Vehicles and Cruise Missiles

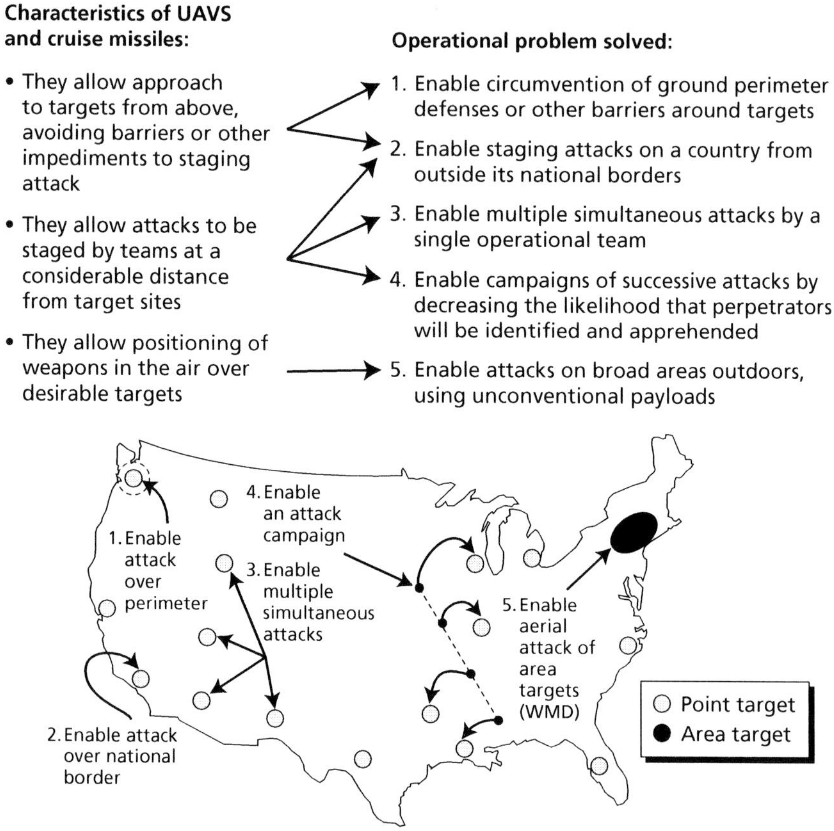

NOTE: WMD = weapon of mass destruction.
RAND *MG626-3.1*

A priori, it appears that the use of UAVs and cruise missiles would be particularly attractive to adversaries that desire these capabilities or those planning operations for which the availability of these capabilities would significantly increase the attacks' impact or their probability of success. The sections below explore each of these capabilities in turn to test this conclusion. They explore both the utility of these capabilities to adversaries and whether alternative attack modes could provide the same (or superior) capabilities at similar or lower cost than UAV and cruise missile systems.

1. Enable Attack over Perimeter Defenses

The vast majority of potential targets of interest to asymmetric attackers lack any perimeter defenses or barriers (Table 3.1), because the United States is a comparatively open society. However, even in the presence of many alternative softer targets, individual protected targets may still be attractive to an adversary if a successful strike on such a target is viewed as particularly valuable in advancing the group's goals. Examples of such targets could include a variety of government sites,

Table 3.1
Target Types and Characteristics

Target Type	Publicly Accessible Locations	Limited-Access Sites	Restricted-Access Sites
Constraints on Adversary:	Direct observation and unopposed assault possible	Direct observation and unopposed or lightly opposed assault often possible	Direct observation and assault denied or difficult
Examples:	**Critical infrastructure:** airports, ports, train stations, bridges, tunnels, hospitals, pipelines, transmission lines, and some dams **Other locations:** national icons, national parks, stadiums, and other large public gatherings, schools, malls	**Critical infrastructure:** nuclear, oil, and gas power-generation facilities, some dams and reservoirs, emergency operation centers **Other locations:** chemical plants, many Department of Defense (DoD) facilities	**Key locations:** selected DoD, intelligence facilities, and political-leadership sites

SOURCE: Target classes and characteristics are adapted from John C. Baker, Beth E. Lachman, David R. Frelinger, Kevin M. O'Connell, Alexander C. Hou, Michael S. Tseng, David T. Orletsky, and Charles W. Yost, *Mapping the Risks: Assessing the Homeland Security Implications of Publicly Available Geospatial Information*, Santa Monica, Calif.: RAND Corporation, MG-142-NGA, 2004.

infrastructure sites, or other public buildings with perimeter security, and mobile defended targets such as motorcades carrying government officials or individual dignitaries during public appearances. There are similarly a variety of targets in the United States that would simply be difficult to attack from the ground, even in the absence of an explicit security perimeter, including such iconic sites as the island-based Statue of Liberty, which are difficult to approach from the ground.

For targets that are protected rather than simply inaccessible, a successful attack must defeat the security perimeter to deliver a weapon to target. The options available for defeating the perimeter will depend on the ultimate objective of the attack. If the target of the attack is stationary and positioned within the security perimeter—e.g., a hardened government installation itself or a secured national icon—then the adversary must penetrate security and deliver a weapon to the target. Doing so necessitates *direct-attack* strategies (Figure 2.2, left) aimed at either getting through, over, or potentially under the security perimeter. In general, attackers have two choices when trying to penetrate a defense directly: by stealth or through the use of force.

If the ultimate target of the attack is mobile—e.g., people inside the perimeter defense or a dignitary protected by a security detail—it might be possible for an adversary to entice the target to leave its protection, thereby making it vulnerable to other attack modes. This goal would be consistent with the use of *indirect attack or challenge-response* strategies (Figure 2.2, middle), which would result in individuals exiting from the protected site. Both modes provide alternative attack strategies that relate directly to the challenge posed by asymmetric attackers from perimeter security around targets and the need to defeat those perimeters to stage successful attacks.

UAVs and cruise missiles could be used in both these modes. Singly, cruise missiles are a direct-attack weapon with the ability to deliver a payload to a target over perimeter defenses. UAVs could be used either to deliver a weapon or in indirect scenarios in which the presence of a threatening air vehicle could be used to attempt to trigger evacuation from or breakdown of perimeter defenses around the target. However, adversaries have a variety of other options that could be applied in both attack modes, many of which are cheaper, less risky, and potentially more effective than UAVs and cruise missiles.

Alternative Means for Defeating Perimeter Defenses

Depending on operational design, a wide range of tactics and weapons is amenable to *direct* attempts to defeat perimeters around desirable targets. If the perimeter can be penetrated by stealth, small-scale weapons (e.g., light emplaced or suicide-triggered explosive devices, basic weapons for use in some armed attack scenarios) can be used. Terrorist organizations have a long history of using such weapons and seeking to smuggle them through defenses, to great effect. Terrorist attacks on the aviation system frequently applied this tactic, and organizations engaged in long-term campaigns of terrorism often used this strategy to neutralize defenses deployed against them. For example, the Provisional Irish Republican Army (PIRA) circumvented some perimeter defenses in this way by bringing bomb-making materials through perimeters in pieces and assembling the devices inside the defended zone (Barzilay, 1975, p. 207).

Stealth strategies allow operatives or weapons entry to a defended zone while avoiding the potential complications associated with fighting through the defenses in a direct assault. Stealth attacks risk complete failure if their secrecy is compromised—for example, through chance detection by a guard force. However, if stealth is maintained, the attacks can be very successful, since defenders will become aware of the attack only when the weapons detonate.

Reliance on stealth can limit the size of weapons for operational design, however. Because larger weapons are frequently more difficult to conceal and defenses can more easily exclude them, stealth strategies frequently employ small payloads that can elude detection (small bombs or carefully shielded devices), or use of delivery mechanisms or personnel that can circumvent security measures or screening. This can mean that, against defended targets, reliance on stealthy modes of attack can force trade-offs between operational risk and the size of the payload used in the attack.

Brute force can also be used to breach defenses by simply overwhelming them. Very large, ground-based (usually vehicle) bombs have been a frequent choice of asymmetric attackers to break through perimeter defenses. The Khobar Towers attack against U.S. troops in Saudi Arabia featured a tanker truck full of explosives. This mode produced a large enough explosion to destroy a building, even though it was stopped outside the security wall. Direct assault by an attack team can also breach defenses.

For ground assaults to penetrate a perimeter defense, teams of attackers must be sufficiently trained and equipped to overwhelm defensive measures. An example of the application of this approach is the al Qaeda assaults on U.S. diplomatic facilities in Saudi Arabia, wherein attack teams successfully penetrated defensive measures to attack employees inside the sites ("Al Qaeda-Linked Group Takes Credit for Saudi Attack," 2004). The attacks on the U.S. Embassies in Africa in 1998 also provide examples of the challenges in breaching perimeter defenses through direct assault. In both cases, preferred entries to the embassy compounds were denied the attackers: In one, this denial led to detonation of the vehicle outside the perimeter; in the other, small arms and a flash grenade were used to breach a perimeter defense (see a discussion of the attacks in FAS Intelligence Resource Program, 1999).

However, breaching defenses has a number of problems, and the outcome of an attack by a small attack team can be difficult to predict. First, time is not on the side of the attacker. The defender will bring an increasing number of forces to the conflict over time, whereas the attacker's strength will only decrease as casualties mount and ammunition is expended. Second, if the target can move, time will allow the target to flee, move into a strongpoint, or otherwise change position in such a way that the attacker will have much greater uncertainty about where the target is located. Finally, direct assaults to breach defenses are uncertain things: Even one additional defender in an unexpected location can hold the attacker off or tip the balance fairly dramatically.[1]

---

[1] In simulations conducted for other analyses for the Department of Defense, we have observed that single gunners have dramatically altered the overall outcomes of engagements when small numbers of ground elements are involved.

While, on balance, it would appear that stealth has advantages over direct assault on defended targets (an advantage reflected in the history of terrorist operational design choices), the risks associated with both could make alternative attack modes attractive. Eluding defenses, by going around, over, or under them, could be appealing, assuming that their benefit in reduced risk could be gained without too much additional cost in time, effort, or other resources. In this regard, direct- and indirect-fire weapons might provide ways to neutralize the defensive benefits of a perimeter. For any large facility, line-of-sight considerations tend to prevent direct-fire weapons from having any utility against interior targets. Non–line-of-sight weapons, such as rockets or mortars, are also useful for attacking defended targets. Examples of groups that have made extensive use of this approach include the PIRA, which used mortars extensively to attack defended security and government targets (Jackson, 2005b, pp. 93–140), and Palestinian terrorist groups using rockets to attack over the security barrier being constructed around Israel (see, for example, the discussions in Richardson, 2002, 2005).[2] Insurgent groups have also used mortars in attacks on defended targets in Iraq, such as strikes on the heavily defended Green Zone. Such systems can carry only small payloads (frequently relying on fragmentation to cause damage to soft or personnel targets) and require that the target location be known with at least some precision—so that the terrorist knows where to aim the weapons—to have a chance of striking anything other than area targets with a significant chance of success. Aerial delivery of significant-sized payloads can be done using aerial vehicles, as was demonstrated in the attacks of September 11, 2001, although doing so requires availability of the necessary equipment and appropriately trained (most likely suicide) operatives to ensure accurate delivery on a target. The Tamil Tigers in Sri Lanka have recently used aerial delivery of explosive weapons from small planes in successful operations similar to traditional military air attacks (David, 2007).

---

2 The use of rockets and missiles by Hezbollah in its 2006 war with Israel is analogous, but we address those attacks in the next section, since they occurred over a national border.

A variety of options is similarly available for an adversary to attempt to bring its target out of a defended perimeter in an *indirect-attack* operation. Frequent approaches for such operations are using initial attacks, planting a bomb, or using a simple armed attack. The results of these attacks attract attention and draw bystanders or security forces out of protected areas and into harm's way. Examples are readily available of nonstate actors in particular utilizing such challenge-response operations, whereby an initial event is triggered to force targets into the path of a second attack. For example, in its attack on the British armed forces at Warrenpoint in 1979, the PIRA staged a bombing that, although it resulted in casualties, was also intended to bring more troops to the scene and enable a larger attack on a higher concentration of service members (Barzilay, 1981, pp. 84–91). Another potential example of such a tactic was the 2002 attack staged in the Bali nightclub district by Jemaah Islamiyah. Two devices were used in quick succession, one inside a bar and another larger device planted in a vehicle on the street outside ("The Bali Bombing Plot," 2002).

Essentially any attack mode could be used as part of an attack scenario that is partially or completely indirect in nature. In such cases, even an ineffective "first attack" that does not penetrate the perimeter (e.g., a bomb that detonates outside the defended zone) might trigger an evacuation or other movement of people that exposes them to subsequent attacks.[3]

A variety of other options could be used as challenges in these types of scenarios. Information operations, such as hoax calls or bomb threats, could be used to trigger an evacuation of a defended location. More-complex provocation operations (e.g., creating a routine emergency, such as a fire, to trigger evacuation and response, triggering of threat-detection systems, or acting to simulate an attack in progress) could also be effective indirect stimuli. The sensitivity of the United States to aerial attack modes since the September 11 terrorist attacks means that an aerial vehicle (whether a UAV or a piloted airplane) clos-

---

[3] This would depend on the specific security policies at the venue targeted for attack. An alternative scenario might be for security to have individuals take shelter in place within the structure, thereby defeating this strategy.

ing into restricted airspace or loitering over a potential target could similarly provoke a reaction leading to evacuation of individuals, thereby neutralizing any protective measures.

Assessment of Options for Defeating Perimeter Defenses

As a mode for attacking a defended target, how do UAVs and cruise missiles compare against available alternatives? While UAVs and cruise missiles have many attractive attributes based on their ability to circumvent defenses oriented to ground threats, they have many limitations as well. From an attacker's point of view, the primary trade-offs would center on the effectiveness at the target (centrally, the nature and size of the payload delivered by the system and the accuracy with which it could be delivered), the probability of getting to the target (chances of successfully breaching the perimeter), and how the costs and risks associated with acquiring and using the relevant technologies needed for each tactical option.

To address the tactical problem posed by a defensive perimeter around a desirable target,[4] the easiest solution for an adversary would be the use of *indirect-attack modes*. If an attack planner is reasonably certain of the behavior a provocation operation or other indirect attack will produce, these operations can neutralize defenses and enable use of almost any attack mode to strike at the target once the target exits the perimeter. Depending on the attack modes a group has available and is experienced in using, the freedom to apply a cheap, familiar, and reliable attack mode could significantly increase the chances of operational success. Among the indirect approaches available, UAVs or other aerial modes appear to be high-cost options relative to, for example, the effect of a telephoned bomb threat or a triggered fire-alarm panel. As a result, unless an adversary has reason to believe that only an aerial provocation will produce the desired evacuation and exit behavior, most other strategies appear preferable.

Adversaries may be skeptical of indirect modes, however, because such modes rely on understanding and predicting the behavior of

---

[4] By *desirable*, we mean a target that is sufficiently attractive to the attacker that choosing another, less-well-defended target is not considered an attractive option.

defenders—and because errors in that understanding could lead to operational failure. In considering the menu of alternatives for *direct attacks*, the attractiveness of UAVs and cruise missiles relative to alternatives would likely be driven by three main questions:

1. Must the attack be staged from outside the perimeter, or can the perimeter be neutralized through stealth or brute force?
2. Is the desired target vulnerable to fragmentation or projectile weapons, or must a significant blast (and therefore a significant payload) be delivered on target?
3. Is the position of the target known with sufficient accuracy, or is man-in-the-loop guidance needed to provide the desired probability for a hit?

Aerial-attack modes in general, and UAV and cruise-missile options in particular, are more complex (and therefore potentially operationally risky) than the simpler, ground-based options.[5] As a result, the choice of such a mode seems unlikely if more-basic tactics can achieve the adversary's goals and the group has sufficient confidence in their own ability to either defeat or overwhelm security measures. UAVs and cruise missiles would appear to compare favorably against alternative technologies only when other options for defeating security perimeters are not available by stealth or by the use of commando teams or large weapons to breach defenses. A preference for simpler, ground-based attack modes is also consistent with data on past terrorist and insurgent offensive operations.

If going over the perimeter seems necessary, available alternatives fall into two classes: small standoff weapons (including mortars, rockets, and UAVs able to carry only small payloads) and larger standoff weapons (including cruise missiles, large UAVs, and attack

---

[5] Depending on the level of experience and capability available to a specific terrorist cell (i.e., are UAVs or cruise missiles an entirely new technology that cell members are learning for the operation?), aerial attack modes may have significant operational risks that would be absent with familiar and more basic tactical choices. Furthermore, such modes would also be more costly than other options if the attackers did not already have access to and experience in flying aircraft. These group-specific factors are discussed more fully below.

aircraft). The functional difference between these classes is in precision and payload capacity: Mortars, rockets, and small UAVs deliver small payloads, generally with low to medium levels of precision, and they rely predominantly on fragmentation to damage the target. More-sophisticated attack platforms can deliver larger payloads (therefore increasing the chance of damaging targets through blast effects) and can do so more accurately.

For larger UAVs, cruise missiles, or attacks involving aircraft to be clearly preferred, the need for precision and/or more payload must be high. If it is not, mortars, rockets, and small UAVs provide simpler attack options at far lower cost (and nonstate groups' general familiarity with mortars and rockets will likely result in a preference for these weapons). Even when precision and large payloads appear necessary, the requirements associated with circumventing perimeter measures do not lead to a clear preference among UAVs, cruise missiles, and aircraft attacks, and choices will presumably be driven by additional cost and risk considerations (e.g., familiarity with systems, availability of pilots [suicide or otherwise] for alternative aerial options).

Different adversaries may have different preferences with respect to the precision required for an attack. While a traditional military planner might demand exact delivery of a payload to a target for a high probability of damaging it, an asymmetric adversary might be satisfied with delivery of ordnance in the general area of a high-profile target—i.e., simply staging the attack might be considered a success, independent of its actual outcome. The value of the higher-level precision afforded by UAVs or cruise-missile systems for simply staging an attack would be much less, presumably making the systems less attractive.

2. Enable Attack over National Borders

Although most terrorist or other asymmetric operations are carried out from short ranges, with weapons delivered directly to the target by suicide operatives or emplaced by an agent who then escapes before detonation, the requirement that operatives be inside a country in order to attack it puts certain limits on the capabilities of attackers. The need for operatives (and potentially weapons) to cross controlled borders and for cells to stage operations in a foreign country produces a level of operational risk and bounds the scale of an asymmetric attack campaign. Consequently, for adversaries seeking to escape these potential constraints, the capability to stage attacks outside the borders of a targeted state could be attractive. Initiating operations outside the country could complicate attribution of the source of the attack or broaden options for intentional misattribution to another actor or nation.

This capability may be particularly attractive for adversaries that have significant technical capabilities and resources outside U.S. national borders (in adversary states, in particular) that are sensitive to even an incremental increase in operational risk involved in moving capabilities through border controls or across lightly defended parts of the national border.[6] An adversary's assumptions about the performance of border security and customs would also shape the attractiveness of attack scenarios beginning outside national borders and about the sites that might be attractive to target in such operations.

Although many small UAVs can operate over only very short ranges, a wide variety of larger systems and cruise missiles have flight ranges reaching into the hundreds or even thousands of miles (Figure 2.1). Such long ranges could make it possible for adversaries to stage attacks from beyond the borders of the United States, circumventing even the qualitative "perimeter" provided by the border security and customs systems that regulate the traffic of people and materiel into the country. As shown in Figure 3.2, even comparatively short-range UAV and cruise-missile systems (with ranges near 100 mi) enable attacks on targets near the coasts from either maritime platforms or from sites very near land borders. Intermediate-range and long-range systems (with ranges exceeding 500 or even 1,000 mi) produce a much larger foot-

6 The success of many actors in bringing materiel and individuals into the country is well known; however, there is some risk of seizure when entering the country. This risk would apply to groups seeking to bring unusual weapons, such as UAVs or cruise missiles, across the border as well. See, for example, Warnakulasuriya (2003).

Figure 3.2
Possible Launch Footprints for Sample Petroleum-Infrastructure Targets for Systems with Ranges of 100, 500, and 1,000 mi

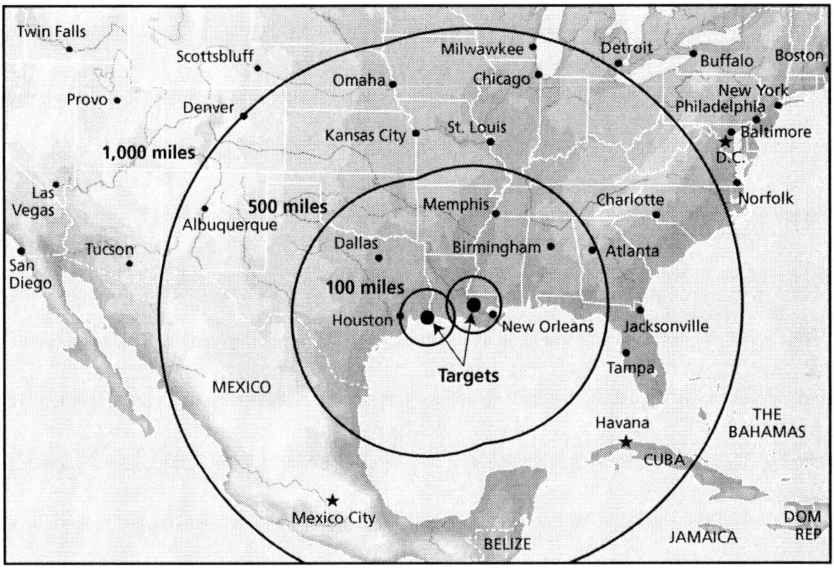

RAND *MG626-3.2*

print around targets from which attacks might be staged. For the target areas illustrated in the figure, which were selected to include high-value petroleum infrastructure sites on the Gulf Coast of the United States, weapons with these ranges could enable the launching of operations from well into Mexico or from a variety of islands in the Caribbean.

Alternative Means for Attacking Across National Borders

Even setting aside traditional symmetric military platforms (e.g., traditional attack aircraft) as options for striking from outside a nation's sovereign territory, options beyond UAVs and cruise missiles are available to the asymmetric attacker. Once an adversary has defined the requirement for attacking from outside the national border, many potential attack modes otherwise attractive to the adversary simply cannot deliver what is required. Ground-based attack modes are essentially irrelevant by definition for attacks across national borders, because of the requirement that the adversary remain outside the country. Among remaining options, the primary requirement is range, to enable the staging of attacks that can reach a relevant number of desirable targets. Short-range systems, such as mortars, many rockets, and snipers, become irrelevant for any target that is not immediately proximal to the border itself.[7]

To attack across a national border, the available attack options are therefore much more limited than are modes that could help an adversary solve the problems posed by a perimeter defense. The majority are aerial modes (UAVs, cruise missiles, rockets of sufficient range to put interior targets at risk,[8] or standard air vehicles used in attack scenarios), although specialized options, such as cyber attack—attempting to cause physical damages to infrastructure systems or other targets through computer-network intrusion, sabotage, or attack—or the use of proxies (e.g., shipping companies[9] or unaffiliated individuals), to carry a weapon across the border and to the target could also be applied.

Traditional, piloted airplanes (in either suicide or nonsuicide operations) provide versatile tools for staging across-border attacks. Established capabilities of organizations to smuggle cargoes (e.g., illegal drugs) across national borders by air demonstrates the viability of this strategy. Airplanes provide significant payload capacities (ranging from car-bomb-equivalent weights of explosives for small planes upward for larger, general aviation aircraft), and direct control by individuals as they cross the national border increases the chance for success if, for example, the vehicle must take evasive action to avoid weather or engage in any protective activities. The costs and risks of using airplanes are therefore less than those for UAVs and cruise missiles with comparable ranges and potentially larger payloads.

Although very different from the other attack modes discussed so far, cyber-attack modes would also provide an adversary with the abil-

---

[7] This is a particularly strong constraint for a nation as large as the United States, in which only a small subset of potentially attractive targets is anywhere near a national border.

[8] Such rockets as those used by Hezbollah in its 2006 conflict with Israel.

[9] Such companies include container-shipping firms and companies providing transport for smaller cargoes via air, ground, or maritime modes.

ity to attack from outside national borders. For effects similar to those produced by the other modes already discussed—actual destruction of a target, injuries, and fatalities, as opposed to simple disruptions of varied kinds—relevant targets for such attacks are more limited, and what the actual capability requirements are for adversaries to cause the desired effects predictably is unclear.

An alternative approach to staging attacks from outside a national border is to use proxies to carry the weapon across the border, allowing the state or nonstate actor to remain outside the border. On a large scale, the potential use of container shipping to move offensive weapons across the border is one manifestation of this strategy. On a smaller scale, other shipping strategies (i.e., the use of "mail bombs") or the use of witting or unwitting individuals to carry weapons could be routes as well.

The value of this approach will be driven by whether sufficient payload can be delivered to a desired target to produce the effects desired by the adversary. Whereas payload capacity may not be an issue for large-scale shipping modes, the amount that could be carried by an individual or shipped in commercial package-delivery systems without raising sufficient suspicion to compromise the operation would be smaller.[10]

Assessment of Options for Attacking Across National Borders

As a mode for staging attacks from outside the national borders of the United States, how do UAVs and cruise missiles compare to available alternatives? All the attack choices described do provide the ability to attack from sufficient distances to overcome the operational problem: Both the aerial modes and the specialized alternatives enable attacks from a sufficient range to occur across a national border. The relative attractiveness of the alternatives will therefore be driven by other characteristics.

---

[10] The Unabomber attack campaign is an internal example of the use of proxies for weapon delivery and the effectiveness of the mode to separate the attacker by a great distance from the site of the attack. The anthrax attack in Washington, D.C., in 2001 is another example.

If the delivery of weapons by proxy is thought capable of producing the desired effects, it is an approach that could be carried out at minimal risk to the adversary group, but with some risk of operational compromise in transit, when the weapon is in the proxy's control. Although cyber attacks could conceivably be initiated from anywhere on the planet, whether they would be attractive to an adversary group would depend on what the group was trying to accomplish in staging an operation. There is consensus that such tactics will likely be able to produce considerable *disruption*; whether they can produce actual *destruction* is more controversial. As a result, the attractiveness of cyber attacks as an option will depend on the group's perception of such an attack's potential effectiveness.

Among the aerial modes, the primary operational characteristic that differentiates piloted planes from UAVs or cruise missiles is the level of terminal control available as the weapon is delivered. UAVs and cruise missiles launched from over a border would, almost by definition, need to have their flight guided through pre-programmed guidance rather than be actively controlled (assuming that the adversary does not have access to satellite communications capabilities for remote control from extreme distances). Such guidance would not allow terminal control for "fine targeting" to address any positional uncertainty in the target or compensate for any dynamic hardening measures (or movement) taken by the target. Plane-based approaches (suicide or not) would provide some terminal control, although nonsuicide strategies relying on autopilot for final delivery to target would resemble a non–terminally guided UAV or cruise missile.

3. Enable Multiple Simultaneous Attacks

For an asymmetric attacker that lacks the capability to carry out large-scale operations or cause damage over wide areas, staging many small operations simultaneously is an attractive strategy for increasing the effect of attack operations. The ability to stage multiple simultaneous attacks has been highlighted as an indicator of terrorist capability, given the coordination and other challenges involved in attacking many targets at the same time (Cragin and Daly, 2004). The recent terrorist-attack operations on the Madrid and London rail systems used this approach, whereby multiple individual explosive devices emplaced in different parts of the systems acted together to produce large-scale terrorist incidents.

The use of coordinated attack operations with multiple devices can also increase the robustness of a terrorist operation. An event that depends on the functioning of a single large explosive will fail completely if that device does not detonate—for any reason. If the operation, instead, involves many independent devices, the attack will still be successful from the terrorists' point of view, even if some fail to go off. For example, the attack on the Madrid Metro system is a tragic example of this reality: Of the 13 bombs planted, three failed to detonate—a failure rate of nearly one in four—yet the operation still produced the carnage intended by its planners (Rojo, 2006).

Just as the range of UAVs and cruise missiles could make it possible to attack from beyond a nation's controlled borders, the ability to "act at a distance" could similarly help in staging multiple simultaneous attack operations. Such medium- to long-range weapons would make it possible for strikes on many targets (even if the strikes are widely separated) to be launched from a single attack point by a small operational cell. Scenarios taking advantage of this operational design could vary from the use of a group of micro-UAVs to small UAVs (e.g., kit-derived remotely controlled planes available from hobby shops) to deliver small payloads to many targets over a limited area, to the use of multiple long-range UAVs or cruise missiles that fan out from a single launch point to strike targets in distant cities across a region or a nation as a whole.

Alternative Means for Staging Multiple Simultaneous Attacks

UAVs and cruise missiles are clearly a viable way for an asymmetric attacker to stage multiple attacks in separate locations simultaneously; nevertheless, recent experience with terrorist activity across the world has shown that there are a variety of other ways to do so as well. Individual bombs can be put in place at each target, with timers set to detonate simultaneously or with remote detonators, all of which can be triggered at once. The attacks on the Madrid rail system are an example of this approach, where the alarm-clock functions of cellular telephones were reportedly used to detonate the bombs (Jane's Terrorism and Insurgency Centre, 2005). A wide variety of other operational designs using emplaced explosives, suicide operatives, and other tactics could similarly be applied to staging multiple asymmetric operations at separated targets.

While many different tactics *could* be used in staging simultaneous strikes, the requirements and risks associated with distinct tactics differ markedly. The number of targets that can be struck using many tactics is limited by the number of operatives an asymmetric attacker has available to stage the operation. Attacks using suicide attackers (whether traveling on foot, in ground vehicles, or in aerial transport, such as general or commercial aviation), armed assaults, snipers, rockets, or mortars require a *minimum* of one operative per site attacked—and, in the suicide operations, the operatives will be lost in the attack. This personnel constraint will limit the maximum number of sites that can be attacked by a cell to its total membership.

Emplacement strategies—leaving weapons at a target for later detonation or attack—make it possible to break out of the constraint imposed by the number of members in the organization. They have been used widely by terrorist groups. Explosive devices are commonly used this way, but mortars and rockets have also been employed as emplaced devices, enabling an operational team to escape or to go on to other activities before the attack itself takes place. While these strategies can enable groups to stage more simultaneous attacks than they have available operatives, groups must devote *time* to gain that advantage—the time required to set up the weapons at all targeted sites—and, as a result of that requirement, take a measure of *risk*.

Unlike attack designs in which operatives may enter the target only once—when the attack is initiated—emplacement operations require entry (and activity) at or near a target significantly before the attack will take place. Security at the target may discover operatives as they are emplacing a device or as they move from target to target, risking the entire operation. Even if the terrorists emplace their devices successfully, operational risks remain: Devices may be discovered before the time of the attack, causing the attack to fail; emplaced devices lack the advantage of a human aiming them during the attack, potentially reducing their chances of successfully achieving the goals of the attacker; and, even if the devices go off, surveillance and forensic evidence is left at the target by the emplacement team, possibly leading to compromise and discovery of the terrorists.[11]

Modes that allow action at a distance can enable groups to break away from the personnel constraint that might limit the number of attacks they could stage, without requiring the compensating increases in the time required and the risks involved in emplacement strategies. These modes substitute *technology* for personnel. Because they do not require entry to the target area before the attack is staged or extended emplacement of the weapons at the target, they also involve less risk of discovery and operational compromise. UAVs and cruise missiles are the most relevant means of interest for this analysis for enabling action at a distance; however, other strategies that enable attack by operatives distant from a target are also relevant.

Of the other modes discussed so far, one example would be cyber attack. Its utility, as described above, would depend on whether the attacking group believed it was an effective mode of attack.[12]

Assessment of Options for Staging Multiple Simultaneous Attacks

In assessing the relative attractiveness of different ways of staging multiple simultaneous attacks, the most important aspect for discriminating between modes would appear to be whether a group feels that its size is an important constraint on its capability. Having person-

---

[11] This issue is explored more fully below, when we examine a terrorist *campaign* scenario.

[12] Theoretically, a group might be able to use proxies to stage multiple simultaneous attacks as well (e.g., shipping weapons to many targets using commercial-shipping or mail systems); however, the ability to *ensure* that the attacks occurred simultaneously would be restricted. Timer- or victim-operated scenarios (e.g., a bomb is triggered on package opening) would require knowing shipping times to a degree of precision to ensure that the weapons would arrive at close to the same times. Remotely detonated scenarios might be viable, assuming that weapons would not be discovered immediately upon delivery (e.g., the group could wait an extended period to be certain that all packages had been delivered to their targets, and then detonate them). All these scenarios involve considerable risks of the operation's being discovered in transit and of the attacks that do occur not happening simultaneously.

nel directly involved in an attack to make adjustments in aimpoint or delivery time at the last moment increases the robustness of an operation and its chances of success. The presence of "human guidance" is a major reason given for the broad use of suicide operations by many terrorist organizations: Having such guidance at the point of detonation enables great flexibility in the final delivery of a weapon to its target (see, for example, the discussion in Hoffman, 2003).

Accepting the tenet that terrorist groups put a premium on success to support their image as effective organizations reinforces continued use of such tactics when they are available. Groups would be pushed away from these modes only when they felt they lacked sufficient group members or, because of difficulty in replenishing their membership, needed to protect their current stock of operatives. Viewed from a domestic perspective, the most important drivers for such a shift would either be difficulties in these groups bringing operatives into the United States (driven by actual or perceived increases in the effectiveness of U.S. border controls) or an inability to recruit group members domestically.

When a terrorist group cannot or will not use man-in-the-loop approaches, whether a group chooses to take on the time and operational risks associated with emplacement strategies or seek out such technological substitutes as UAVs and cruise missiles will depend on its risk perceptions, preferences, and available resources and capabilities. Historically, when terrorist groups have wanted to preserve the safety of their membership, their preferred strategy has been to utilize emplaced bombs or other weapons, suggesting that most groups consider the time costs and operational risks associated with those modes acceptable. The Provisional Irish Republican Army is a prime example of such a group. It put a premium on protecting its members and made extensive use of emplacement strategies for both single-attack and multiple-simultaneous-attack operations.

Many UAVs and cruise missiles could deliver payloads comparable to those for emplacement strategies and would not expose operatives to the same risks. These modes would have specific technical risks of their own that would not arise for more basic attack modes: They are more difficult to operate effectively and a remotely controlled or programmed flight vehicle could have technical or other problems (e.g., with weather conditions). The importance and scale of these risks would depend, in part, on a group's level of experience and expertise with these platforms.

The apparent preference of terrorist organizations for emplacement strategies over such alternatives suggests that, to date, the benefits of technological substitutes, such as remotely guided vehicles,[13] have not been viewed as sufficient to compensate for any significant investments in the resources, time, and training needed to acquire and use them. This situation could change if more-effective defenses are put in place that make detection of emplaced weapons (or individuals emplacing them) more likely, therefore increasing the real or perceived risks associated with those modes.

4. Enable an Attack Campaign

For exerting pressure on a society, extended campaigns of terrorist violence can have a much more sustained effect than single terrorist attacks. Repetition of violence, even if individual attacks are smaller than the macro-scale events of a September 11, 2001–style attack, can maintain psychological pressure and fear in a population and increase the total costs that an adversary can impose on the targeted state.

Examples of such campaigns and their effects are readily available in the repeated attacks by Palestinian terrorist groups on Israel, attacks by the Liberation Tigers of Tamil Eelam (the Tamil Tigers) in Sri Lanka, and the activities of the Provisional Irish Republican Army in Northern Ireland and Great Britain. An example of the effects of a campaign of violence at a much smaller scale can be found in the reactions in the Washington, D.C., metropolitan area to the attacks by

---

[13] Note that ground-based remotely piloted vehicles are also an option. There are reports that PIRA experimented with building vehicles that could be guided by remote control or by the Global Positioning System (see Harnden, 2000, p. 208; Geraghty, 2000, p. 212). This technology has been cited as an example of technical convergence between PIRA and FARC, which has also reportedly experimented with such vehicles (Reuters News Service, 2002, p. A31; and personal interview with a former security forces member, England, March 2004).

the Beltway snipers (John Allen Muhammad and Lee Boyd Malvo) in 2002 (see, for example, Morin and Deane, 2002, p. A01). The strong reaction to these specific cases of murder (ten victims in a year when 474 people were murdered in the Washington, D.C., metropolitan statistical area [Federal Bureau of Investigation, 2002]) emphasizes the potential for a campaign of violent incidents to have a disproportionate reaction from the perspective of creating terror.

Repeated attacks can also undermine public confidence in security forces and law enforcement, can bolster the image of a terrorist organization's effectiveness, and can help to trigger overreaction by authorities, which can magnify the effect of terrorist activities on the targeted state. Extending operations over time also provides an adversary with the opportunity to learn from attack to attack, potentially increasing its effectiveness (Jackson, 2005a; Jackson, Baker, Cragin, Parachini, Trujillo, and Chalk, 2005).

In modern societies, however, violent campaigns can be difficult to sustain. Whereas the characteristics of open Western societies can make it easy for clandestine groups to operate for extended periods without coming to the attention of authorities, once a group has staged the first attack in its campaign, it has clearly announced its presence. The capabilities available in most countries—e.g., surveillance cameras at or on approaches to targeted sites, forensic-evidence analysis, investigative capability—create the potential for group members to be identified and apprehended after an initial attack, particularly after high-profile attacks, to which large amounts of investigative resources can be committed in focused apprehension activities. The speed with which police in the United Kingdom identified suspects in the July 7, 2005, bombings of the London Underground is a prime example of how data collected at the scene of an attack can advance an investigation (House of Commons, 2006). As a campaign continues, and as each additional attack provides new opportunities for evidence collection and witnesses who might identify the perpetrators, sustaining repeated operations can become even more difficult.

The range provided by UAVs and cruise missiles and the difficulty in tracking their flight paths could be valuable to an asymmetric adversary carrying out a campaign within the United States. Simply enabling an attack team to stage far away from the intended target of the attack shields cell members from surveillance or human observation, reducing the chances of identification and apprehension. The potentially long time of flight to the target also would provide the attacker a window for escape, further separating participating individuals from evidence of participation in the launch.

In this case, the ability to act from a distance could limit evidence collection and foreclose many of the more straightforward routes to link perpetrators to the attack operation. The ability to separate attacker and target can also allow the staging of an attack from one law-enforcement jurisdiction, state, or region of the country into another, possibly taking advantage of organizational conflicts or dysfunctions to reduce the effectiveness of investigation of past attacks and to complicate efforts to heighten vigilance for future attacks. All these factors could reduce the chances of apprehension before, during, or after an operation and, therefore, help ensure that operatives remain free to stage the next attack in their campaign.

Alternative Means for Sustaining an Attack Campaign

Just as terrorist organizations have used attack modes other than UAVs and cruise missiles to stage simultaneous attacks, groups have staged offensive campaigns with a wide variety and many combinations of weapons. Many terrorist organizations, ranging from traditional ethnonationalist groups, such as the Provisional Irish Republican Army, to more-contemporary Islamist groups, such as al Qaeda in Iraq, have carried out campaigns of operations using conventional explosive weapons and small arms: emplaced bombs, vehicle bombs, and their suicide analogs, and firearms and infantry weapons in armed assault–type attacks. For sustaining repeated attacks over time, smaller-scale weapons (e.g., emplaced bombs over vehicle bombs) can be advantageous both because of the logistical issues in producing sufficient materiel to carry out large-scale attacks repeatedly[14] and because it is easier to hide and clandestinely transport small weapons than large ones.

---

[14] For example, when PIRA routinely used large-scale vehicle bombs in its bombing campaign, the amount of explosives consumed was substantial. Estimates by some observers of

While these ground-based attack modes provide great flexibility in payload and employment, all require direct entry to the target, with the associated risk that the attackers will be identified and the operational security of the cell broken. Even suicide operations,[15] in which escape by the individual operatives is not part of the plan, have some risks: Trace back of the suicide operative's path through surveillance or investigation can compromise the remainder of the cell. Campaigns of such operations can also be personnel-limited in the same way as efforts to stage many simultaneous attacks, unless the group can replenish its personnel through recruitment or reinforcement.

Standoff blast and fragmentation weapons, such as mortars and rockets, reduce operational security issues to some extent by enabling attack from near, rather than directly at, a target. The separation of the launch point for mortars and rockets can also provide a delay before defenders at the target can get close to the site where the attack was actually carried out, enabling escape of attackers or the use of secondary charges to destroy forensic evidence. Use of timed launch of emplaced weapons can further reduce operational-security concerns by allowing operatives to escape before attack initiation.

PIRA, a group that made extensive use of mortars in its multiyear campaign, took advantage of all these approaches. The advantages come at a price, however. In most cases, rockets and mortars can deliver only comparatively small explosive charges to the target, limiting their potential impact.[16] Sniping tactics, in which firing points can be comparatively easily concealed and difficult to detect even if observed, can similarly provide operational-security benefits that facilitate a campaign of attacks, as the experience of Washington, D.C., in 2002 demonstrated. However, the potential effect of any individual sniping attack

---

the consumption rate in large bombing operations went as high as two tons of explosives over a period of three weeks (Geraghty, 2000, p. 45).

15 Whether carried out via an individual-carried bomb, vehicle bomb, or even an aerial suicide attack (in a general-aviation or commercial aircraft).

16 Exceptions to this generalization have been observed. PIRA produced very large "barracks-buster" mortars that could throw much larger charges, but only over comparatively short ranges.

is limited compared with what many terrorist organizations seek in designing their operations, making them potentially unattractive as an alternative, in spite of their operational-security advantages.

Assessment of Options for Sustaining an Attack Campaign

For sustaining an attack campaign within the continental United States, UAVs and cruise missiles appear comparatively attractive, in spite of the expense and technical risks associated with them. Because of the capacity of domestic (local, state, and federal) law-enforcement organizations, attack modes that require penetration of targets seem unlikely to allow extended campaigns, particularly once a few attacks have occurred and security forces (and the public) are on heightened alert. Standoff alternatives are better, although evidence left at attack points and public or law-enforcement efforts to detect terrorists setting mortar baseplates or launchers for follow-on attacks could compromise the attack cell. Sniping tactics are sufficiently limited in capability that even their desirable operational-security characteristics will make them viable for only a subset of potential terrorist operations.

Although there are payload limitations on what UAVs and cruise missiles can carry, this is less of a comparative disadvantage with respect to an attack campaign than with other asymmetric operations: The premium placed on operational security would likely drive alternatives to small scales as well—e.g., use of explosive devices comparable to or smaller in size than the payload of many UAV systems. Furthermore, alternative ways to maintain such a large separation between the launch point (the attack team) and the target are not readily apparent, making these systems stand out from the perspective of an attacker seeking to repeat attacks in the face of investigative pressure and heightened vigilance.

Among available UAVs and cruise missiles, the premium placed on limiting the launch and operational signatures of the systems will push toward smaller rather than larger weapons. Once one attack utilizing an air vehicle has been staged, authorities and the public will be looking for "terrorists launching planes or missiles." As a result, maintaining operational security would be easier with smaller rather than larger vehicles. Clandestine transport of large weapons (e.g., cruise-missile systems or UAVs at the high end of the size spectrum) would be more difficult and would require more-extensive preparation and staging before launch. In addition, cruise missiles in particular normally use rockets with significant burn times, which would produce significant visual and acoustic signatures that could compromise the operational team.

In contrast, small UAVs have key advantages. Many can use soft launch (using hand-launched takeoff for the smallest and pneumatic or powered takeoff for the rest), and many models have engines quieted for surveillance missions. As a result, these systems appear to have a particularly interesting niche for an attacker seeking to sustain an asymmetric campaign.

5. Enable Aerial Attack of Area Targets with Unconventional Weapons

For an adversary that has acquired and intends to use an unconventional weapon, whether chemical, biological, radiological, or nuclear, effective delivery of that weapon will presumably be a major factor in the operational design for an attack. The past behavior of nonstate groups and other adversaries has suggested that, for most groups, an unconventional weapon payload will be a high-value asset. Despite broad concern about the damage these weapons could potentially produce, there have, fortunately, been few instances of groups' acquiring or effectively using unconventional weapons. This fact suggests that a group that did acquire enough weaponized agent to create a mass-casualty incident would put a premium on the success of the operation (so as not to waste a high-value group asset) and on delivering the weapon in such a way to ensure broad enough exposure to individuals or a targeted area to cause damage and disruption through contamination.

The ability to stage attacks from the air could be attractive for achieving these goals, particularly when the intended target is an area rather than a point target: Release from above can enable a threat agent to be dispersed in ways[17] to increase the chances of affecting targeted individuals or widening the area contaminated in the attack. Similarly, control of detonation altitude for a nuclear weapon is part of what must be considered about the scale and type of damage and fallout it will create. As a result, unlike most high-explosive weapons, dispersal of a chemical, biological, or radiological agent or detonation of a nuclear weapon *above* a target is a viable and potentially preferential attack mode. UAVs and cruise missiles could be attractive ways to disperse such agents because of the flexibility they provide for positioning the unconventional payload over a wide range of potentially attractive target locations.

Alternative Means for Dispersing Weapons over Area Targets

Whereas UAVs or cruise missiles would allow aerial dispersal of unconventional agents over a wide range of targets, a variety of other potential release modes is also available to potential adversaries.[18] Although aerial modes of delivery have advantages, an asymmetric attacker may be satisfied to deliver its agent on the ground. Previous terrorist use of these weapons, such as Aum Shinrikyo's use of sarin on the Tokyo subway, has demonstrated that these weapons can be deployed by operatives that simply deliver them to their intended targets. For larger operations, a spray truck could be employed. The use of the mail to disseminate the anthrax used in the 2001 incidents in the United States is another example of the use of a much more limited, ground-based delivery mode.

As discussed in previous sections, ground-based modes can enable the use of very large payloads (e.g., up to the scale of a large chemical tanker truck for an industrial chemical used as a chemical weapon) and the advantages associated with a man-in-the-loop operation. Building on the assumption that a weaponized unconventional payload will be a

---

[17] For example, a line release of an agent rather than a point release.

[18] Explicit consideration of alternative modes of dispersal prevents the analyst from conflating the attractiveness of the unconventional weapon (which terrorist groups have expressed intense desire to acquire, if not the broad technical capability to acquire) with the attractiveness of UAVs and cruise missiles as a delivery mode for that weapon.

high-value asset to an asymmetric attacker, we can see that these strategies also have the advantage of maintaining direct control over the weapon until delivery. The primary disadvantage of these strategies is that they place the agent at ground level and are limited in how much they can spread the weapon (relevant for chemical, biological, and radiological) compared with aerial-release modes.[19] A less conspicuous approach might be to disperse toxins through the air-intake system of a building.

Mortars and rockets, in addition to providing the advantages associated with standoff attack options, can provide aerial dispersal of an agent above a target, assuming that they have been designed to detonate at altitude rather than on impact. They could provide a delivery mode for dispersible chemical, biological, or radiological agents, although they would not be relevant for nuclear weapons. This strategy may also have some technical risks for a nonstate actor, associated with producing and perfecting the use of mortars and rockets in this mode. The primary disadvantages are accuracy concerns (potentially increasing the chance of operational failure if the weapon does not hit the intended target), the comparatively small volume of agent that each such weapon could be used to deliver (capping the scale of an attack they could be used to launch), and the fact that each detonation will produce a point release of agent over a relatively small area.

If a desired target is in an area in which a terrorist could "get above it" without using a vehicle—e.g., a site in a city near tall buildings or an area with geographical features at different elevations—a group could gain the advantages of aerial dispersal without sacrificing any of the advantages of ground-based modes. The weapon could be kept under control of members of the group until release, but dispersal from a high point would allow broader spread of the agent. This strategy could similarly be used for a nuclear device, and selection of the desired altitude could simply require identifying on which floor of a skyscraper the device should be placed for the attack.

---

[19] They may also face operational risks associated with security measures at the desired target, as described in the discussion of UAVs and cruise missiles as a strategy for circumventing perimeter defenses.

A potentially viable strategy, use of an elevated location is only a true alternative for the subset of targets proximal to such sites. If a group was flexible about the targets it was attacking, it is possible that an organization might select a site according to the availability of such a dispersal location. Otherwise, the attractiveness of this route would be defined by the local geography and architecture around the group's desired target.

In addition to UAVs and cruise missiles as aerial modes, standard piloted planes represent an attractive alternative for delivery of unconventional weapons. Small or general-aviation planes provide flexibility in determining where an agent will be delivered, and they have the advantage that the weapon is kept under group control until the attack is carried out. Some planes have significant payload capacities, enabling larger-scale operations if desired by groups. Agents could also be employed in line releases from these platforms, making a broader spread of an attack possible. However, to utilize this strategy, groups will have to gain access to aircraft (through purchase, rental, or theft) and to qualified pilots who can fly them to the target location.

Assessment of Options for Dispersing Weapons over Area Targets

There are several key elements for discriminating among the alternative ways of staging unconventional attacks. An adversary using an unconventional weapon is assumed to put a premium on attack success: Depending on the adversary, the unconventional payload could be of very high value (certainly the case for nuclear weapons, likely for biological, less so for chemical or radiological). In such scenarios, success depends on the ability to deliver an appropriate volume of the agent to target (which clearly varies among different agents, targets, and attack scenarios) and the level of certainty that the payload will be delivered successfully.

The higher the perceived value of the payload is, the more likely groups will want to maintain direct control over the weapon until the point of delivery. Most of the alternative attack modes provide this control. The tightest control is achieved with ground-based suicide tactics, dispersal from elevated locations, and the use of piloted aircraft. UAVs and cruise missiles do not offer such direct control. As a result, for a high-value payload (particularly a nuclear weapon and biological agent), an adversary would have to have a high level of "trust" in the ability of the UAV or cruise missile to deliver the weapon to the target. Any technical uncertainty associated with the functioning or use of the system, as well as a variety of factors outside the group's control (e.g., weather conditions) could undermine that trust and push organizations toward using more-robust or fault-tolerant approaches to deliver their weapons.

Choices between ground and aerial modes (where all *aerial modes*—UAVs, cruise missiles, general aviation, elevated locations, mortars and rockets—are more complex operations than simply ground-based delivery) will be driven by the perceived benefit associated with putting the agent above the target. Whereas a group that is seeking to optimize the damaging outcomes of its attack might always see those advantages, a group that is satisfied with the potential outcomes of a ground release might never see the justification for the added complexity of adopting an aerial-dispersal mode. Ground modes also generally allow delivery of larger volumes of material than can many aerial systems. Magnitude could provide an additional perceived benefit of the more basic tactical designs for groups that either wanted to use more agent to create larger attacks or simply to increase the probability of operational success.

If an aerial dispersal is preferred, what would drive choices among UAVs, cruise missiles, and piloted aircraft is less clear, beyond the concern about unmanned systems' technical and other operational risks discussed previously. Payloads for many of these systems can be significant (with the exception of UAVs at the low end of those systems' capabilities ranges), although lower than can be carried by many ground-based forms. Piloted airplanes (over pre-scripted flight for cruise missiles or UAVs) would enable a terminal-guidance capability that could adapt to behavior or countermeasures at the target. This capability suggests that UAVs and cruise missiles do not provide markedly different capabilities for aerial dispersal over the use of standard piloted airplanes, but they could provide an alternative for groups that were constrained in their access to pilots or to planes.[20]

Conclusions

What conclusions will state or nonstate groups reach regarding the attractiveness of UAVs and cruise missiles compared with available alternatives? To guide thinking on this question, we examined five operational problems that these systems appeared most useful for solving and assessed them against other possible solutions in areas in which

---

[20] Our conclusions are comparable to those reached in a 2001 National Intelligence Estimate, which approached the related but distinct topic of ballistic-missile threats and the use of those systems for unconventional attack. After citing cruise missiles as a more likely threat than ballistic-missile systems, the study also cited a number of much lower-technology vectors for unconventional weapons delivery and highlighted their potential advantages over intercontinental ballistic missiles (ICBMs) and (to a lesser extent) cruise missiles (National Intelligence Council, 2001):

> [C]oncern remains about options for delivering WMD to the United States without missiles by state and nonstate actors. Ships, trucks, airplanes, and other means may be used. In fact, the Intelligence Community judges that U.S. territory is more likely to be attacked with WMD using nonmissile means, primarily because such means:
>
> – Are less expensive than developing and producing ICBMs.
>
> – Can be covertly developed and employed; the source of the weapon could be masked in an attempt to evade retaliation.
>
> – Probably would be more reliable than ICBMs that have not completed rigorous testing and validation programs.
>
> – Probably would be much more accurate than emerging ICBMs over the next 15 years.
>
> – Probably would be more effective for disseminating biological warfare agent than a ballistic missile.
>
> – Would avoid missile defenses.

The differential in cost, required capability, reliability, accuracy, and effectiveness between UAVs and the ground-based means cited in the report would be smaller than between them and ICBMs or cruise missiles. However, comparable arguments would apply and suggest that there would be barriers to the attractiveness of these systems for unconventional-weapon dispersal.

aerial capability seemed to have the advantage. By viewing these systems in their best light, our goal was to explore what advantages adversaries might gain by pursuing them.

In general, most of these comparisons resulted in the conclusion that UAVs and cruise missiles were certainly viable and potentially effective attack modes. However, they frequently did not appear to have major advantages over other ways of carrying out operations against similar targets. Where they did appear preferable, the choice for these systems was driven by the actions of the defense or security measures that were in place: Defenses foreclosed the alternative attack modes or concerns about a potentially compromised plan pushed the attacking group farther away from its desired targets. The advantages had a price, however: greater complexity, technological uncertainty, cost, and risks associated with these platforms. As such, UAV and cruise missiles appear to represent a "niche threat"—potentially making some contribution to the overall asymmetric and terrorist threat, rather than being an attack mode likely to be widely embraced by such actors.

How much UAV and cruise-missile systems contribute to the overall threat faced by the United States is further determined by whether or not asymmetric attackers will actually face (or believe that they face) the operational problems these systems are most useful for solving:

- Although aerial systems could be useful for *circumventing perimeter defenses*, most targets in the United States lack any such defenses, making many targets readily accessible without a need for such capabilities.
- While *attacking from outside the national borders* could be attractive, currently U.S. national borders are viewed as comparatively porous to both people and materiel, and key materials can be found domestically. Therefore, adversaries may not see the need to initiate their attacks outside those borders. Given a decision to attack from outside the national borders, attractive alternatives to UAVs or cruise missiles include manned aircraft, the use of proxies (containers or shipping services) to carry weapons across the border, or (very different from the other attack modes) cyber attack.

- The value of using these types of technologies to *stage multiple simultaneous attacks* is largely driven by groups constrained either by an unwillingness to plant many explosive charges (and risk their discovery) or having a limited number of operatives available for emplacement or suicide operations.[21] Even if groups operating in the United States are limited in their ability to recruit individuals domestically, the perceived problems with U.S. border control suggest that they may not face substantial barriers to bringing operatives from abroad.
- The effectiveness of domestic law enforcement suggests that *sustaining protracted terrorist campaigns* in the United States could be difficult. Nevertheless, examples of serial bombers (e.g., the Unabomber) and other criminals (e.g., the Washington snipers) indicate that campaigns of some length can be carried out. Such historical examples may reduce the likelihood that adversaries will see campaign sustainment as a concern meriting this type of specific technological approach.
- Although *dispersal of unconventional weapons* over a densely populated area target is of concern, the wide variety of ways for dispersal limit the chances that this issue will be viewed by an adversary as a problem requiring a UAV or cruise missile to solve. Even so, in this scenario it is the adversary's possession of the unconventional payload, not the delivery mode, that is the primary source of the threat.

Consequently, even for operational problems for which UAVs and cruise missiles appear potentially attractive, if adversaries do not believe that they face those problems or that those problems materially constrain their capabilities, incentives for pursuit of these weapons may be further weakened.

Moreover, in the event that groups do choose these UAVs and cruise missiles as delivery systems in attack operations, the likely *outcome* of representative operations where they could be used does not

---

[21] Recent attacks in Madrid, London, and Mumbai suggest that such constraints have not affected contemporary groups' attack planning.

differ significantly from many alternative ways of staging those attacks. Use of UAVs and cruise missiles generally forces the use of smaller payloads, comparable to the low-to-mid-range capability of the current terrorist arsenal.

In scenarios for which the attractiveness of UAVs in particular derives from the desire to launch an attack at long range because of security concerns, payloads are constrained even further by the capacities of the relevant systems, resulting in weapons with capabilities at the low end of current systems widely used by these organizations. The reality of payload capacities and range suggests that these systems have built-in limits that bound the scale of the threat they pose when they are constrained to conventional payloads.[22]

---

[22] A potential exception to this general statement would involve attacks on specific targets for which the impact of an attack is not linearly constrained by payload size—for example, targets that have built-in failure modes that might be initiated by a small attack but produce a larger total impact such as some industrial or infrastructure targets. Similarly, the use of unconventional payloads, such as a contagious biological, could also be an exception, because the spread of the disease would provide an amplification mechanism that could compensate for a small initial payload size.

CHAPTER FOUR

What Are the Terrorist Group Characteristics and Preferences Relevant to the Acquisition and Use of Technology?

Comparing the capabilities and applications of different attack technologies is just one part of an effort to anticipate an adversary's choices and future tactics, and the uses that technologies can be put to are only one driver of organizational decisionmaking. The preceding discussion has focused largely on the likely benefits of UAVs, cruise missiles, and potential alternative-attack technologies. Important as well are the costs and risks that organizations must bear when they choose one technology or tactic over another.

In contrast to the broader conclusions that can be drawn about technological capabilities, assessing how these costs and risks might shape decisionmaking must be approached largely on an adversary-by-adversary basis: The following individual-group characteristics and preferences can shape tactical and operational decisionmaking and constrain the opportunities available to a specific organization:

- access to and costs associated with UAV and cruise-missile technologies
- access to and costs associated with alternative technologies
- ability and willingness to develop the expertise necessary to operate the systems
- technological preferences.

Each is discussed in turn in the following sections.

Access to and Costs Associated with UAV and Cruise-Missile Technologies

Even if an asymmetric attacker is interested in using UAVs or cruise missiles, it cannot use these technologies if it cannot get them. For obtaining systems, adversaries have two choices: (1) obtaining them from someone else (e.g., through transfer from a state or other legitimate owner of the system) or (2) developing a system for themselves. Access to technologies will therefore differ from group to group according to differences in the group's or its members' past experience that shape the ability to access technologies through either of these routes.

Whether an adversary can obtain a system from someone else depends entirely on whether the group or state has the right connections to technology sources and can exploit them. At one level, this can be viewed as a "yes or no" question: If a group has the right linkages, it is possible; if it does not, it is not possible. From a more detailed perspective, however, this can be viewed as a question of cost: Whether the group has access to technology sources that will offer technology at a cost the group is willing and able to pay. In this case, costs can be viewed in a number of ways beyond simply financial consideration.

The most obvious conduit for such systems, particularly the higher-end military systems, to hostile nations or nonstate groups is via a supportive nation-state that believes it can use that actor as a proxy to launch attacks that are in its interests, or to simply support an organization because of philosophical, political, or even religious sympathies. If a group has such a sponsor and that sponsor is willing to provide these technologies freely, then the apparent cost of their acquisition and use to the group could approach zero, thereby strongly favoring their usage.

While such transfers are clearly possible, transferring states may have concerns about the ability to trace the source of technology, limiting the chances of providing such aid anonymously.

The story for commercial UAVs, ranging from more sophisticated models to simple hobbyist systems, could be quite different, especially as UAVs become more ubiquitous. In many cases, such systems could be provided essentially anonymously because of their broad availability, either in the United States or abroad.

An adversary's capability to develop technologies for itself will depend on the technical skills and expertise it can bring to bear on the problem. These skills and expertise will differ markedly among different classes of adversaries. It is one thing for nation-states such as Iran, with its broad resources and other advantages, to develop cruise missiles or field sophisticated UAV systems. It is another for a much smaller entity, such as a terrorist organization, to develop similar capabilities on its own. Different UAV and cruise-missile applications have different requirements, however. More-basic modification of existing systems or improvisation of similar UAVs to provide weapons carriers would have fewer technical requirements.

Whether or not a specific group could overcome either the lower technical barriers to acquiring basic systems or the more-formidable ones for sophisticated military-style systems would depend on its past technical experience and availability of complementary skills and capabilities. Acquisition of UAVs and cruise missiles by groups with experience in similar systems would be easier than for groups pursuing them as new technologies from scratch. An effort to develop technologies internally will similarly have resource requirements ranging from willingness to invest the time and effort of the groups' technical experts to the financial outlays required to procure needed inputs for the process. Such decisions also involve opportunity costs, because the efforts and resources applied cannot be used for other purposes.

The choice to pursue UAVs or cruise missiles from external sources or to develop them inside the group has its own operational-security costs and associated risks. Reaching outside the group requires involving actors over which the group has no control and who could betray them. In contrast, an internal effort will involve many activities that could be detected by security organizations and compromise the group. It will also require maintaining secrecy over the longer time period required to develop and build the systems.

Access to and Costs Associated with Alternative Technologies

Similarly, the arguments made about access to and costs of UAVs and cruise missiles made previously apply to all of the alternative technologies that adversaries could use to stage attacks. Any alternatives that are wholly unavailable to a group would be factored out of consideration, potentially increasing the relative attractiveness of UAVs and cruise missiles.

For alternatives that are accessible to a group already, the cost of using those familiar attack modes may be low, particularly when compared with the costs of acquiring a new technology or tactic such as cruise missiles or unmanned aerial vehicles. Using them may also pose fewer operational-security risks to the group than an effort to obtain and master a new technology. For alternatives that a group does not possess, the costs of obtaining them compared with UAVs or cruise missiles would shape their relative attractiveness. The case of cyberattack capabilities, cited briefly in the preceding chapter, is a particularly good example. Because such capabilities would be unusual for many asymmetric adversaries, their costs would likely appear high to most groups.

The perceived level of operational and other risk associated with alternatives can also be a significant contributor to their apparent cost as compared with novel technologies such as UAVs and cruise missiles. From an adversary's perspective, the risks of failure and other adverse outcomes may be much less with familiar tactics and technologies: a benefit of using the tried and true over the new and unproven. Group preferences and risk tolerance will shape the effect of this risk cost on decisionmaking. Depending on the adversary, "success" and "failure" could also mean very different things. For example, different groups have defined operational success as

- *Staging the Operation*—frequently termed an attack to "wave the flag" and demonstrate the ability to act, sometimes simply carrying out an attack may be considered success, independent of any particular outcome.

- *Achieving a Threshold Level of Tactical Effect*—for a specific attack, the adversary may have some minimum amount of damage it believes is required for the attack to contribute to its goals. For adversaries that are seeking to manage escalation (i.e., they do not want the asymmetric attack to stimulate significant escalation on the part of the United States), there may also be a maximal threshold output below which they seek to remain.
- *Maximizing Tactical Effect*—a maximizing adversary may seek to simply cause as many fatalities and/or as much damage as possible for a given attack.

If simply "waving the flag" is not enough and specific outcomes are desired from an operation, the risks associated with new technologies could reduce their perceived attractiveness. Conversely, if those technologies are seen as essential for achieving threshold or maximizing effects, this perception could provide added incentives to pursue them by reinforcing their perceived benefits.

Ability and Willingness to Develop the Expertise Necessary to Operate the Systems

Whether the group already knows how to use UAVs and cruise missiles could shape its decision to pursue them as an attack option. As discussed above, lack of needed expertise can significantly increase the risk of operational failure and could be particularly problematic for complex technologies that differ considerably from those a group customarily uses. As a result, if a group does not already know how to operate and employ UAVs or cruise missiles effectively, it will need to learn how to do so.

Such learning takes resources, however, which could produce a disincentive for pursuing the systems in the first place. Learning efforts require time and manpower, and they also risk loss (or intentional expenditure) of systems during training activities, unless the group has access to safe havens for carrying out such training, learning efforts may involve overt actions that might be detected by intelligence or law-enforcement organizations and compromise the group's operational security (see the discussion in Jackson, 2005a; Jackson et al., 2005).

Technological Preferences

Beyond issues of technological access and capability that can shape the technology decisions of individual groups in particular ways, simple preferences—what a group likes and dislikes—will shape those decisions as well.

A group might see an inherent value in pursuing high-technology weapons—for example, to bolster the group's reputation as a sophisticated and potent adversary in the eyes of its enemies or sympathizer communities. In this case, the inherent preference will shape judgments about the value of the technology in spite of the ability of alternative attack modes to deliver comparable operational effectiveness.

Conversely, a group that sees itself as traditional or established in particular modes of operation could have inherent disincentives to pursue new attack types if those attack types are viewed as departures from the status quo or incompatible with the group's current operations.

Conclusions: Two Decisionmaking Pathways

When analysis focuses only on the potential operational benefits of UAVs and cruise missiles and of alternative attack modes, it is possible to objectively compare different components of effectiveness and reach relatively general conclusions about the attractiveness of different systems in given attack scenarios. Such conclusions can be supported by physics and engineering analysis that weighs weapon effects and related variables that shape attack outcomes.

In contrast, the effect of organizational characteristics on technology and tactical choices is not governed by objective and universal factors; nonetheless, it is important for assessing the threat. It is possible that broad shifts in the environments or preferences of many adversaries could make these weapon systems appear more attractive. For example, if a state sponsor or other source of technology began providing these systems freely to many adversaries, the apparent costs of these systems would go down across the board.

In the absence of trends that broadly influence the perceived costs of these and alternative attack modes, groups' expertise in using these systems and their technology preferences will have an effect that will be inherently idiosyncratic, driven by the specific details of individual groups, their members, their histories, and their preferences. Such factors make it more difficult to generalize about likely bottom-line effects on adversary behavior and could lead to episodic adoption of these attack modes by individual groups, even when their performance appears inferior to alternative ways of staging operations.[1]

In an effort to bring together the instrumental decisionmaking process of the red analyses of alternatives described in the Chapter Three with the group-specific influences on decisionmaking described here, we found it useful to consider a two-path model for adversary decisionmaking. Both paths, or approaches, can be described by simplifying an adversary's tactical and operational planning to two decisions—(1) choosing an attack mode or technology and (2) choosing a target—and distinguishing two paths for adversary decisionmakers based on which of these two decisions they make first. Some organizations may have a preference for one path over another; other organizations may take both paths simultaneously.

Examples of both types of planning processes can be identified in the history of state warfare and the violent activities of nonstate actors. For states, an example of the first path is development of extensive catalogs of targets vulnerable to current capabilities (e.g., targets in states of concern that could be readily struck via airpower). Examples of the second path can be found in special-operations planning, wherein

---

[1] This possibility is consistent with the observation that terrorist groups frequently do not seek to *optimize* the performance of their operations (where *performance* could be viewed from a number of different perspectives, as discussed earlier with respect to definitions of successful operations). Frequently, "good enough"—*satisficing* rather than *optimizing*—is sufficient.

high-value goals are identified, and then teams and capability sets are assembled or developed to enable operations to achieve them.

Both behaviors are observed in terrorist organizations as well. Groups with significant experience and capability in specific weapons (armed attack or explosives use, for example) may drive their target selection toward sites or individuals vulnerable to those attack modes. In contrast, groups also identify specific, high-profile targets they believe are particularly desirable (e.g., the assassination of prominent individuals or the destruction of particularly symbolic sites) and then design operations specific to striking those targets.[2] These two options are framed as the two paths an adversary might take, depending on which factor has a dominant influence over attack planning (Figure 4.1).

Path I

On Path I, the adversary chooses the technology first. For the purposes of this analysis, this choice would be of a UAV or a cruise missile (CM) as the

Figure 4.1
Two-Path Model for Adversary Decisionmaking

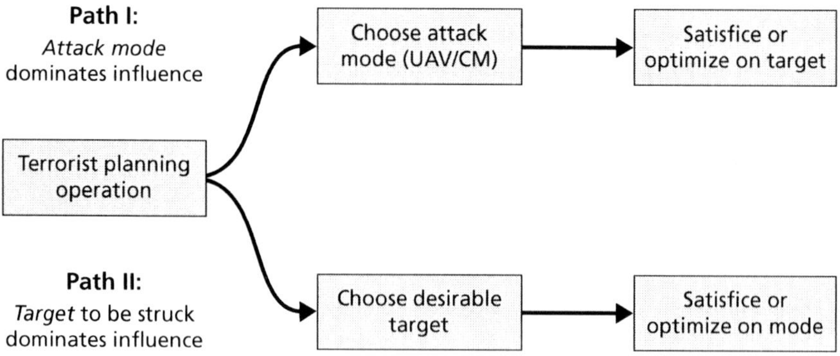

RAND *MG626-4.1*

[2] For a discussion of these two modes of operation in a group that exhibited both types of decisionmaking behavior, see Jackson (2005b, pp. 93–140).

preferred attack mode, because of the group-specific factors discussed in this chapter. Such an initial choice could be made simply because the technology is available: A state sponsor provided the group with training and a weapon and, as a result, the apparent cost of its use is small. Alternatively, the group may have a particular affinity for the attack form, given the perceived benefits of the choice.

Furthermore, as the analysis of the five operational problems in Chapter Three demonstrated, specific tactical concerns can produce this sort of decision as well. Such a technology-first choice echoes the outcome of our analysis of UAVs in particular as campaign enablers, whereby the desire by a group to carry out a string of attacks could make the technology attractive and the specific targets that would be attacked during that campaign had limited or no effect on the outcome of the technology comparison. The outcome of our examination of these systems for attacks across national borders could produce a clear preference for UAV or cruise-missile technologies, provided specific operational risk and other concerns made some alternatives unacceptable and an adversary constrained itself to attacking from outside the country.

Path II

On Path II, the primary consideration is what to attack, not how to attack it. As a result, the target to attack is chosen first and alternative attack modes are assessed according to their perceived cost and effectiveness for attacking such a target. In our previous analyses, the case most clearly demonstrating this type of thinking was for attacking a target already protected by a perimeter defense. The decision to attack a defended target implied a specific focus on that identified target (given the variety of potential substitute targets available), and the focus of analysis was how different alternative attack modes compared. To a lesser extent, the scenario examining dispersal of an unconventional weapon over an outdoor target was similar, although more potential attractive substitute targets are available for such attacks. In these cases, comparison of the strengths and limitations of available modes seldom left UAVs or cruise missiles as clear winners among the alternatives.

These two broad classes of adversary behavior enable bringing together the comparatively clear technical differences between these weapons and alternative attack modes with the more subjective and idiosyncratic differences among groups that may affect their tactical choices. Combining the two classes provides a structured way for considering both the lessons that an examination of the full menu of attack modes available to asymmetric adversaries provides for focusing defensive efforts and the complications that differences in adversary preferences and decisionmaking create for building a prudent defensive strategy.

CHAPTER FIVE

Considering Defensive Strategies and Options

Our examination of the attractiveness of UAVs and cruise missiles for attacks in the homeland has significant good news for defense planners. The functional capabilities of these platforms are comparable or inferior to many options currently available to terrorist groups; thus, no strong incentives exist for widespread adoption of UAVs and cruise missiles. In most cases, the level of threat to most targets of concern was driven by the payload delivered to the target, whether conventional or unconventional, not by the fact that these systems allowed delivery of that payload from the air. As our analysis of alternatives suggests, under some circumstances, delivery of a weapon by UAV or cruise missile may actually be less effective and more risky than available alternatives, and adoption of these modes by adversaries might actually be preferred by the defense.

UAVs and cruise missiles as attack systems currently do not appear to be a major threat faced by the United States; however, it would not be appropriate to entirely dismiss these weapons. Even in the current security environment, in which many targets in the United States are vulnerable to simpler and cheaper attack modes, these more specialized systems look attractive under certain circumstances. In particular, our examination of UAVs as enablers of attack campaigns inside the country demonstrated a set of circumstances—particularly when the survivability of the attack team is valued—under which the benefits of these systems could make them attractive. In addition, targets with defensive perimeters may push attackers to aerial attack options.

Group technology preferences—or the willingness of states to provide these groups with UAV or cruise-missile technologies—could similarly push groups to use these systems, even when alternative attack forms would be comparable or superior. As a result, cruise missiles and UAVs could represent a niche threat, even if they are unlikely to become major elements in these groups' operational planning in the near term.

Shifts in the domestic security environment going forward could also change the relative attractiveness of attack modes, making these unmanned aerial systems more desirable. For example, while it is currently assumed that the effectiveness of U.S. Border and Customs controls is not such that groups would face major constraints in bringing operatives and materiel into the country, significant improvements in the effectiveness of these controls could make UAVs and cruise missiles more attractive because they enable attack from outside the border. The combination of these factors challenges security planners to craft a prudent approach to a potential, although currently limited, threat, given many competing demands for resources inside and outside security applications.

Just as our red analyses of alternatives suggested a range of ways in which specific operational goals could be achieved, a similar analysis from the defender's perspective suggests a variety of defensive approaches, each with differing costs and anticipated benefits. Keeping in mind the multiple options available to the attacker, the defense cannot afford to focus on one attack mode in isolation. Instead, the defense must strive to develop capabilities that are effective against multiple threats, investing only in defenses tailored to a specific threat when the expected benefit markedly outweighs the expected costs.

The remainder of this chapter focuses on identifying (1) low-cost, high-benefit defense options to address the cruise-missile and UAV threats, (2) broader defense investments against multiple threats, and (3) defense recommendations based on the implications of these two analytic exercises.

Options Available to the Defender

As in our broad assessment of attackers' options, an inclusive view of defensive responses explicitly seeks to get away from stovepiped approaches to this type of problem. In stovepiped approaches, a single defensive measure or a small number of options are considered in isolation, and potentially valuable strategies could be overlooked.

In crafting a defensive strategy against the threat posed by UAVs and cruise missiles, there is a temptation to begin by examining active-defense systems designed to shoot down these threats. Examining such options in isolation addresses only a small portion of the defense options to counter these systems. The defender has a variety of options and is not constrained to focus on the very short timeframe in which an aerial attack occurs (Figure 5.1). Instead, the defender should consider defensive options along a broad timeline, depending on when in an adversary's activities the options are expected to exert their effects.

As this chapter shows, not only will the defense tend to favor investments to stop an attack before it occurs, but, for defending against cruise missiles and UAVs, such investments enjoy significant advantages over investments in active defenses.

Identifying and Catching the Perpetrators: Intelligence, Law Enforcement, and Forensics

Many activities are intended to uncover terrorist attacks before they occur. Counterterrorist intelligence,[1] law-enforcement activity,[2] and

---

[1] In our analysis, we have defined *intelligence activities* as those efforts carried out by national-, state-, and local-level police and intelligence agencies that are specifically focused on detecting terrorist activities and on monitoring the activities of terrorist groups.

[2] *Law-enforcement activities* are the general activities of police at the state and local levels, not necessarily focused on terrorism, whereby routine interaction between police and individuals may lead to the discovery and disruption of a terrorist plot. In the maritime domain, the Coast Guard's and others' efforts to monitor and regulate ship contents and traffic would fall into this category.

Figure 5.1
Timeline for Defensive Options Against UAVs and Cruise Missiles, Arrayed Against Attacker Activities

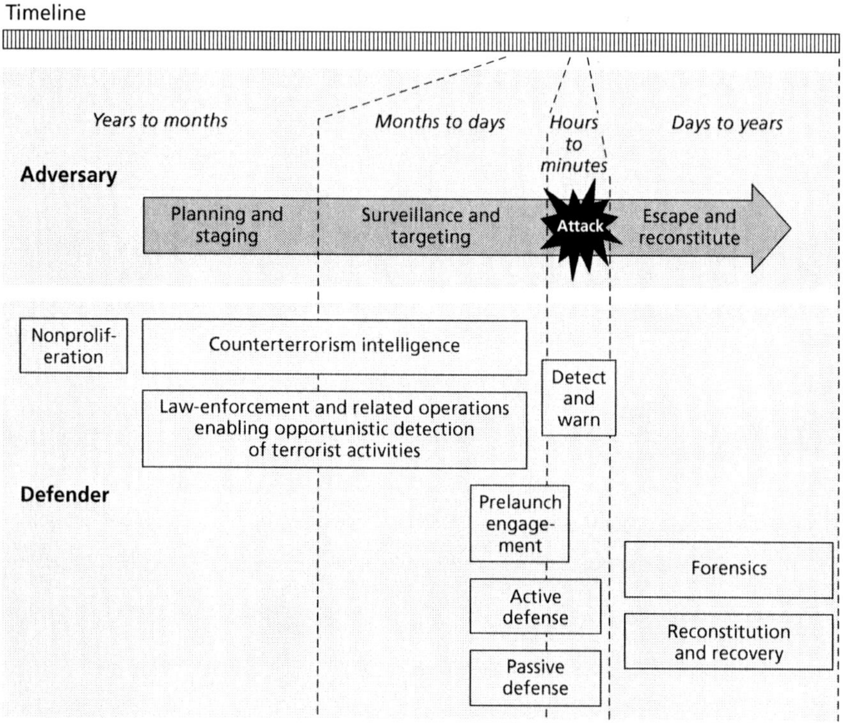

RAND *MG626-5.1*

forensics investigation[3] all focus on identifying and catching the perpetrators either before or after an attack is carried out. In contrast to the active-defense approaches aimed at neutralizing UAVs or cruise missiles in flight, these efforts focus on chasing the terrorists, not the weapons.

The potential for terrorist use of UAVs and cruise missiles may require some adjustments in the way these activities are currently carried out or the development of some new technologies—e.g., explicitly adding activities involving remotely controlled planes to lists of

[3] *Forensics investigation* is viewed as the techniques and capabilities designed to assist in identifying the perpetrators of an attack after it has occurred.

behavior for police to watch for, educating domestic sellers of these technologies to be on the lookout for and to report suspicious purchases, more-focused monitoring of the transfer of these technologies internationally, and forensics techniques specifically designed to mine for information the remnants of a UAV used in an attack.

In general, efforts focused on these functions will address this threat even without specialized investments—as part of overall efforts to counter terrorist activities, detect and investigate crime, and so on. As a result, even if they are stimulated by concern about the risk of UAV and cruise-missile use, investments in this area will produce other benefits even if these specific threats do not arise.

These approaches also have the advantage of pushing back defensive action into a region where timelines are longer (Figure 5.1). This focus on the long lead time before an attack may also have the advantage of attacking the threat at a point where plots are more vulnerable.[4]

Examination of the five operational problems for which UAVs or cruise missiles might be particularly attractive does indicate some particular areas where investments would improve the ability to address these threats. The flight ranges of UAVs and cruise missiles, which figured significantly in the ability of these systems to facilitate attack campaigns, strike many targets simultaneously, and attack from outside the national borders, increase the importance of effective coordination and information sharing across law-enforcement jurisdictions and among intelligence agencies. If an attack occurs using a long-range system, the launch point could be in a wide circle of jurisdictions around the target site, necessitating the involvement of many organizations in the search for the perpetrators. Similarly, a group carrying out an attack

---

[4] A scenario involving an attack in which cruise missiles or UAVs are released from a maritime platform outside the national borders is instructive in this case. Once the attack platforms are loaded onto the ship, an attacker must endure an extended wait time before the ship will move into an appropriate position for weapon release. Actions that have the potential to identify the ships carrying the attack platform, including the full range of cargo security, Coast Guard, and other maritime domain-awareness activities, have that entire time to work. If and when these security activities identify a ship as threatening, the adversary has few options for responding, making it likely the operation will fail entirely.

campaign could move from place to place as they launched, stepping from one jurisdiction to another as the attacks continued.

In the wake of an attack, forensic capability is needed to mine the attack scene for information and identify the perpetrators, potentially breaking the perpetrators' ability to carry out additional attacks. Given that the separation of the attacker from the target distinguishes this from other attack options, a forensic capability could be particularly important for these threats.

Many forensic capabilities useful in this context would also contribute to other investigations, so investments in this area could produce benefits in areas unrelated to this threat. However, investments specific to these threats also have a role to play. Collection of relevant technology and information to support the development of better forensic approaches—e.g., acquisition and study of foreign UAV and cruise-missile systems in ongoing efforts to gather and exploit technical intelligence—would have an important role in building the foundations needed for post-attack study and for determining any unique signatures of a specific state's systems.

Furthermore, although standard forensic approaches for assessing what remains at an attack scene are clearly important (e.g., recovery of trace evidence from the attack vehicle), other, more specialized technologies not traditionally viewed as "forensic" are also relevant. For example, systems for tracking aerial targets may be useful for forensic purposes, even if their performance is not good enough to support detection, warning, and active defense (discussed below).[5]

Controlling the Spread of UAVs and Cruise Missiles: Counterproliferation

In contrast to the short time windows in which defensive approaches focused on detection and response must operate, efforts to keep weap-

5 For example, in a scenario in which a ship offshore launches a weapon, a tracing system that did not have high enough resolution to allow immediate identification of the source vessel could still contribute to follow-up investigation. The population of vessels within a reasonable launch area could provide a starting point for the investigation; if launches occur at later times near different populations of ships, the system could help identify the source ship by a process of elimination.

ons out of the hands of adversary groups have a long, open-ended timeframe for action. Success in such efforts is becoming more difficult, however. The increasing availability of UAV and cruise-missile systems is making counterproliferation increasingly difficult. At the low end of the spectrum—i.e., hobbyist vehicles and above—these approaches are essentially no longer relevant since such systems are so readily available and entirely uncontrolled.

For high-end systems, including large-payload and long-range UAVs and military missile systems, counterproliferation efforts are still relevant and have the benefit of attempting to broadly affect many groups' capabilities simultaneously. For example, the Missile Technology Control Regime is intended to constrain the transfer of missiles, including cruise missiles and UAVs with a range greater than 300 km, and particularly those with payloads greater than 500 kg. Although the MTCR is certainly not a perfect instrument, diplomatic efforts to strengthen its transfer protocols focused on long-range, large-payload weapons could constrain the availability of some of the most destructive systems.

To address the specific concerns that groups might use UAVs or cruise missiles for disseminating unconventional weapons, counterproliferation is similarly relevant. However, such efforts should be aimed at the weapons, not the delivery system. Given alternative modes available to groups for using such weapons, the subset of attack scenarios that involve UAVs and cruise missiles is not the primary source of the threat; rather, the possession of unconventional weapons by such groups is.

Enabling Targeted Sites to React Before Impact: Detection and Warning

Simply *detecting and providing some warning* of an incoming UAV or cruise missile could provide the opportunity to react and limit the potential for harm. Just as indirect-attack modes rely on poorly crafted reactions to an assumed threat to bring individuals or vehicles into a vulnerable position at which they can be attacked, intelligent reactions to an actual threat can significantly reduce its effectiveness. Moving a dignitary to a defended position, making the target safe (e.g., triggering a chemical-plant shutdown to minimize the effect of impact), and either evacuating occupants from a vulnerable building or moving them to appropriate shelter in place all represent ways in which warning time could be used to limit the damage from these types of weapons. A detection-and-warning strategy could be implemented broadly (i.e., a detection network that covered the entire country or every major metropolitan area) or could be focused locally, protecting specific targets of concern.

Investments in detection and warning must overcome two fundamental hurdles: the difficulty of detecting and identifying these threats, and the need, once these threats are detected, of a functional warning system that can provide information on what actions should be taken. Many UAV and cruise-missile systems are small and are not readily picked up by the radar systems that cover much of the country for such routine applications as monitoring air traffic. Even when they are detected, clear monitoring is needed to plot their paths and identify likely targets. And when a target is identified, if a system does not provide enough data on the incoming vehicle to assess its size and potential capabilities, triggering an appropriate response may be difficult: If the incoming vehicle is small, evacuating occupants of a building may increase rather than decrease the risk to them. In contrast, if the vehicle's payload is large, the attack could trigger collapse of the structure, making sheltering in place a greater risk. Understanding the size and maximum lethal impacts of the attacking platform would help decisionmakers take better-calculated risks: of prompt evacuation, of taking additional time to sweep evacuation routes and only then evacuate, or of sheltering in place.

Operational timelines provide major challenges: Systems would have to be in place to transmit information rapidly enough from the location of the detection platform to any identified targets so that decisions could be made on response actions, and reliably enough that the system would not be scrapped because of frequent false alarms, which would be too disruptive of daily life to sustain.

Acting Against the Incoming Weapon or the Launcher: Active Defenses and Prelaunch Engagement

Active-defense strategies attempt to shoot down the UAV or cruise missile in flight before it reaches or as it is reaching its target. The United States already fields active air-defense systems to defend deployed military forces from air attack by fixed-wing and cruise-missile threats. Air defense of the homeland differs from air defense of fielded forces because the area to be defended is much greater, unintended consequences may be more severe, and the identification of threats is much more challenging.

Active defenses must operate within a relatively short period, particularly if an adversary chooses to launch its UAV or cruise missile close to its intended target, shrinking flight times to a span of seconds. This situation results in requirements for rapid and sure detection, identification, and engagement of the target.

Current air-defense systems are able to defend only relatively small areas from such a threat as UAVs or cruise missiles, which can fly at low altitudes and use the contours of the earth to make it difficult for sensors to detect them.

Even if cruise missiles and UAVs are detected, they still need to be identified accurately before defenders can take action: In the homeland, deploying a weapon requires a very high level of confidence that the threat was correctly identified so that an air-defense system does not mistakenly shoot down an aircraft. For instance, an airplane carrying a governor accidentally strayed into restricted air space over the nation's capital. An active-defense system is challenged to distinguish between this errant pilot and a hostile attack. An active-defense system would also involve the risk of unintended damage from engagements, depending over where the intercept occurred. So the defense not only has the burden of certainty about what it shoots down but also about where any debris or weapons used in such an intercept will fall.

The necessity to be certain about the identity of an aircraft and to calculate fallout and unintended damage from an intercept will introduce significant delays in the reaction time of a defense system. That any active-defense system will have some delay time between detection and engagement of a target provides a straightforward way for an adversary to evade the effects of the defense system. Giving up some of the benefits of distance by launching an attack close enough to a target that the total flight time of the weapon is less than the required time envelope in which the defense can respond, an adversary can neutralize the effect of a defense on weapon effectiveness. Even command, control, and communications delays on the order of minutes (e.g., five to 15 minutes through the entire process) would mean that the active-defense system could never be effective against attacks initiated within sufficiently short flight times of desired targets. Eliminating the possibility for such outside-the-envelope attacks would require maintaining ground exclusion zones around targets. Such zones are potentially possible for some isolated installations; however, they would never be practical for large numbers of targets in urban or populated areas.[6]

Prelaunch engagement strategies attempt to detect individuals setting up to launch a UAV or cruise missile and engage them or the weapon before it is launched. Therefore, this approach requires broad detection capabilities, although it must detect threat activity on the ground rather than in the air. This detection must be effectively coupled with the ability to respond and engage individuals across *wide geographic areas* (defined as a radius around relevant targets of the maximum flight range of UAVs and cruise missiles). These practical requirements make these approaches at least as sensitive to outside-the-envelope attacks as do active-defense options.

Strengthening Targets to Survive Attack: Passive Defenses

Passive defenses encompass a wide range of investments to harden a target against attack. They do not require detection and response capabilities, because they are designed to reduce the chances of successful

---

6 The potential for outside-the-envelope attacks means that it is physically impossible for many active-response architectures to address the full threat-space posed by these systems. The potential for such attacks is particularly problematic for the threat of UAVs and cruise missiles, since the requirements for an adversary to determine the boundaries of a system's performance would be quite straightforward. Staging repeated attacks at different distances from targets would enable an adversary to determine the response time for the defense system and, from then on, to stage all attacks outside its envelope of response—essentially eliminating the value of the system against this threat.

attack or limit damage whether or not the target is aware that an attack is under way. Some passive defenses are relatively specific for the threat posed by UAVs or cruise missiles (e.g., barriers, such as catch nets or fencing, to interfere with the weapons' approach to a target); others could provide much broader benefits (e.g., hardening or resilience measures that could protect against blast hazards or even some effects of natural events) across terrorist-related or even larger threats. In contrast with the previous strategies, passive defenses can be implemented only on a target-by-target basis, although wide implementation at many sites could provide the equivalent of a national- or regional-scale effort.

Bouncing Back from Attack: Response, Recovery, and Reconstitution

In the wake of a terrorist attack, the capability to halt damage from the attack and recover from that damage can significantly reduce the total damage that an adversary can inflict. The need for this capability is particularly clear for infrastructure or other targets for which damages caused at the site may deny valuable functions that magnify the total cost of the event over time. As a result, the ability to recover and reconstitute after an attack is an element of an overall defense.

We define *response, recovery, and reconstitution measures* as including the capacity for medical treatment to address human casualties and the ability to quickly repair damages and reestablish the functioning of a targeted infrastructure system. Investments in these capabilities can be applicable to diverse threats and to the damages caused by natural events as well.

When considering an attack using an unmanned aerial vehicle or cruise missile, we think that it is also relevant to note that the reaction of policymakers and political leadership after an attack could play a role in shaping the public's reactions to and, consequently, the overall effect of such an attack. Attacks at the lower end of the UAV capability spectrum, such as one using a commonly available hobbyist model airplane to deliver an explosive payload to a target, could best illustrate how such reactions could shape the impact of an attack.

The outcomes of an attack using a model airplane would be comparable to an attack using a similar amount of explosive in either a suicide bombing or as an emplaced bomb, or they might even be less serious, given the challenges associated with staging such an attack. After an attack using such an aerial vehicle, one could imagine a range of possible reactions by political leaders. At one extreme, an emphasis on the novelty of the attack—"an entirely new threat using an unmanned aerial vehicle"—could actually advance the goals of an asymmetric adversary by heightening the significance attached to the operation and increasing the level of terror in the population. At the other extreme, an emphasis on what is commonplace or even inferior about the operation—"our enemies have been reduced to attacking us with remote control planes"—could instead contribute to reducing the effects of the operation and reducing the attractiveness of such attack modes over the longer term.

Since the primary effect of many asymmetric operations is the terror and psychological reactions they produce, understanding how actions and statements after the fact could either magnify or help to diminish those reactions are an important—and essentially no-cost—element of response and recovery.

Comparing the Options: Bases for a Blue Analysis of Alternatives

Now that we have laid out a variety of options that could contribute to a defensive approach to UAVs and cruise missiles, we need to ask, How should defenders make decisions among them? For exploring defensive decisionmaking, a conceptual blue analysis of alternatives is useful to think through the similarities and differences among the options and assess how decisions could be made. We have framed this discussion as a conceptual cost-benefit analysis,[7] looking first at the benefits of vari-

---

[7] A quantitative cost-benefit analysis of the varied approaches of defending against these threats, and the varied options that could be used for such defense, was beyond the scope of this work. Our cost-benefit analysis should be better viewed as a qualitative discussion of different strategies' costs and benefits to inform thinking about these issues rather than identify a single preferred or optimal decision outcome, similar to the discussion of attackers' options in Chapters Two and Three.

ous options, then at their costs, then risks that could affect whether the apparent benefits of the defenses are realized in practice.

Our examination addressed the following questions:

- How do the defense options differ in their effect on the threat from UAVs and cruise missiles?
- Do the options provide defensive benefits with respect to other forms of attack beyond UAVs and cruise missiles?
- How do the costs compare?
- Are the solutions appropriate for the homeland?
- Are there technical or organizational challenges that might threaten the benefits of an option's being realized?

How Do the Options Differ in Their Effect on the Threat from UAVs and Cruise Missiles?

Since the central goal of implementing a UAV/cruise missile defense would be to provide protection from these threats, the most fundamental element of an analysis of alternatives is a comparison of the effectiveness of different defense options. Direct comparisons among the full range of options are difficult, since the way in which each option approaches providing protection is significantly different. Some approaches, including active defenses, prelaunch engagement, nonproliferation, intelligence, and law enforcement, seek to neutralize the threat entirely, either by denying adversaries what they need to stage these attacks or by preventing them from staging them. Others, such as passive defenses or detection and warning, seek to blunt the effect of an attack when it occurs, rather than prevent it. Finally, response, recovery, and reconstitution and forensic capabilities do not seek to prevent an attack at all; they seek to address only the consequences of an attack's occurring and enable apprehension of the perpetrators to limit the risk of additional attacks.

Another key variable affecting the benefits that a defensive measure delivers is the geographic scope of the protection it provides. Investments in some defensive measures provide general protection that covers the entire country. The scope of different defensive options is diagramed in Figure 5.2, along the horizontal axis of the conceptual graph.

Investments in law-enforcement capabilities, intelligence, forensic capabilities, efforts at prelaunch engagement, and nonproliferation are examples of strategies for which investments provide, or at least attempt to provide, general protection across the nation. Other defensive approaches, most notably, passive defenses, are inherently site-specific, and a single implementation protects a single target. Passive-defensive efforts can be scaled up by installation of similar defenses at many potential targets, thereby spreading protection over a wider area; however, the costs of doing so will increase linearly with the number of targets protected.Three other defensive options—response, recovery, and reconstitution; detection and warning; and active defense—could

Figure 5.2
Defensive Approaches and Their Scope Against the Asymmetric Threat from UAVs and Cruise Missiles, for Comparison

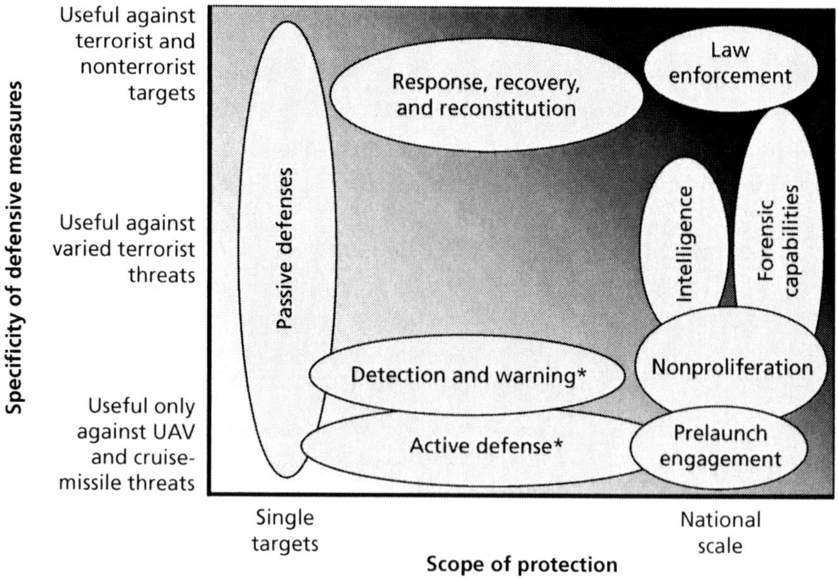

*The scope of protection depends on the scale of implementation.
RAND *MG626-5.2*

similarly provide protection over a range of areas from individual targets to broad national implementations. Costs (discussed below) increase as the protected area expands, although not as clearly and linearly as those of passive hardening.

Do the Options Provide Defensive Benefits with Respect to Other Forms of Attack Beyond UAVs and Cruise Missiles?

Rather than thinking about how a particular defense affects a single threat, it is better to assess how defensive options affect the overall threat to the nation across the full range of ways an adversary might choose to attack. As suggested by our red analysis of alternatives, many attackers will not focus only on a single attack option. So, considering defenses for only "one attack at a time" may skew decisionmaking: The defense may successfully prevent one type of attack but leave other options wide open. In general, the desired outcome of defensive investments is to reduce the total threat to the nation, not just to foreclose individual types of attack.

How to reduce the total threat is a particularly relevant question for UAVs and cruise missiles, since these systems are fundamentally delivery vehicles for other, damaging payloads. Even assuming that active defenses or other measures prevent an adversary group from using UAVs or cruise missiles, the group is still left with its conventional or unconventional payload and many alternative ways to make use of it. As a result, the net effect of even a successful defensive effort may simply be displacement to different attack means or to different targets and, therefore, little actual improvement in national security.

The range of benefits from various defensive options across different threats is summarized in Figure 5.2, on the vertical axis of the conceptual graph. Some defensive options are highly focused on the threat of UAVs and cruise missiles (e.g., focused nonproliferation efforts, detection and warning systems, attempts at prelaunch engagement, many active-defense options, and some passive defenses). For example, high fencing or catch nets that seek to protect a target from aerial approach (a passive defense) or a detection system focused on small aerial targets (relevant to detection/warning, prelaunch engagement, and active defense) may provide little in the way of benefits for anything other than UAVs or cruise missiles. Other defensive options are more general, applying not only to use of these weapons but to other terrorist or asymmetric actions (e.g., counterterrorism intelligence, forensics, and some types of passive defenses, such as general blast hardening). The remaining defenses are even more general, applying not only to terrorism or UAV/cruise-missile risks but also to addressing many other threats from natural or man-made sources.

How Do the Costs Compare?

Estimates of rough orders of magnitude for costs of different defense options are shown and described in Table 5.1 to enable qualitative comparisons of the resource levels involved in making different choices.

Some of the defensive options discussed in the preceding sections involve additional incremental expenditures to supplement already-existing activities to also address threats from UAVs and cruise missiles. For example, the changes (if any) that need to be made to intelligence and law-enforcement activities to ensure that UAV and cruise-missile threats are covered and that sufficient additional resources are available to address them would likely be comparatively modest. Counterproliferation activities already address some of these threats (notably, missile technologies) and similarly might require only modest increases in support to cover a larger portion of these systems.

The costs for passive defenses vary across a wide range, from very inexpensive measures on a per-target basis (e.g., fencing and catch nets) to large investments in robustness and resilience of major infrastructures. This range makes it more difficult to generalize about the costs for that class of defenses. In contrast to most other options, most of the costs associated with response, recovery, and reconstitution are not paid up front (with the exception of the preparedness investments needed for rapid-response activities) but are paid to repair the damages caused after an attack occurs.

By far the most expensive options are those that seek to detect and respond to UAVs or cruise missiles themselves. The magnitude is driven by the costs of detection systems sufficiently advanced and sensitive to detect the weapons (ground-based or aerial radars, human

Table 5.1
Qualitative Order-of-Magnitude Costs of Defensive Options, for Comparison

Defensive option	Requirements: Nonproliferation program	CT intelligence effort	LE vigilance and reporting	National coverage detection	Regional (e.g., metro area) detection	Detection at individual targets	Engage and kill—national	Engage and kill—metro area	Engage and kill—individual target	Harden a target to these aerial attacks	National forensic capability for attribution	Resources to fix damages if attack occurs
Nonproliferation	$											
CT intelligence		$										
LE and related operations			$									
Prelaunch engagement		$	$	$$$$			$$$$					
Detection and warning												
National				$$$$								
Regional (per city)					$$$							
Location-specific (per target)						$						
Active response												
National				$$$$			$$$$$					
Regional (per city)					$$			$$$				
Location-specific (per target)						$			$$			
Passive defense												
Location-specific (per target)										$-$$$$		
Forensics			$								$$	
Reconstitution and recovery												
Repair costs (per target)												$$

$ Millions of dollars
$$ 10s of millions of dollars
$$$ 100s of millions of dollars
$$$$ Billions of dollars
$$$$$ 10s of billions of dollars

NOTES: CT = counterterrorism; LE = law enforcement.

spotters, and so on), command and control systems to trigger either warning or an active defense, and appropriate platforms (e.g., aircraft, missile systems, or other weapons) to close with and engage either the UAV or cruise missile in flight or to seek to engage the individuals launching them on the ground rapidly enough to prevent their escape. For these defensive options, the large numbers of both detection sites and defense installations (airbases, missile sites, firing points, etc.) needed to protect any significant fraction of the country (with a response time fast enough for success) result in costs that rapidly increase into many billions of dollars.

Are the Solutions Appropriate for the Homeland?

Many potential defense investments against cruise missiles and UAVs will originate in the Department of Defense. Employing a system in the homeland that was designed for use in a combat zone will require a review of whether such a system is suited to such a task. Although a defensive measure may appear beneficial if its acquisition and operations/maintenance costs are compared to its estimated benefits, its benefits may never be realized if there are barriers to the defense actually being used. Some defensive approaches are not affected by this risk: Passive defenses function without being triggered and so are always in use. Others are not.

Active-defense and prelaunch-engagement options could have significant barriers to use, since potentially deadly force or weapons that could produce unintended damage would be involved. The potential for damage exists whether the defense succeeds or fails. A surface-to-air or air-to-air missile that misses its target could fall into a populated area, causing unintended damage. However, even a successful intercept will bring down the UAV or cruise missile and its payload, perhaps preventing them from hitting the intended target but possibly causing them to hit something else. For instance, in the First Gulf War, U.S. Patriot missiles successfully intercepted an Iraqi SCUD missile launched against Riyadh, Saudi Arabia, but the debris killed one person and wounded 23.[8]

Even detection and warning strategies could have significant barriers to use, although not as serious as those affecting active-defense strategies, because the triggering of evacuations or other actions at

[8] Department of the Navy (1991).

potential targets is not without cost. There will always be a disincentive to issuing warnings, particularly if the warning needs to be issued to a general geographic area (e.g., a threat heading toward the Chicago metropolitan area) containing many possible targets.

Are There Technical or Organizational Challenges That Might Threaten the Benefits of an Option's Being Realized?

The final critical question in comparing defensive options is, "Will the defenses actually perform when the nation calls on them?" In our discussion of adversary decisionmaking, simple technologies and tactics were highlighted as advantageous because they could be more robust and fault-tolerant than more-complex methods—more likely to actually work when they were used. Similar arguments apply to defensive measures. As systems become more complex, success relies on more elements performing appropriately and in a coordinated fashion for a successful outcome.

Many passive-defensive options are the simplest of technologies: Fortifications for blast hardening or barriers to prevent an aerial approach to targets have no moving parts and therefore little risk that they will not function during an attack. In contrast, active-defense systems rely on many sequential activities—an entire kill-chain from the detection of a potentially hostile target through confirmation and engagement—that must be carried through or the system will fail. Any breakdown or delay at any stage, whether in transmission of information, decisionmaking, finding and fixing the target, or executing an attack, could allow the attacker to reach its target and undermine the value of all components of the defensive system.

Defense Conclusions: Choosing Among Available Options

Looking across the variety of options available, we now ask, "How should defense planners allocate resources to address the threat of asymmetric use of UAVs and cruise missiles?" Our analysis suggests that, although UAVs and cruise missiles are potentially attractive to asymmetric adversaries in some cases, they are not so disproportionately. It is never possible to say that such groups will not broadly take up a particular technology; however, in this case it appears that, even if they do, it will not dramatically change the level of destructive power such groups can bring to bear when compared to available alternatives. As a result, while prudence suggests putting some defensive measures in place to address the threat, their scale must be small, commensurate with the level of threat these systems pose.

The primary UAV and cruise-missile attack scenarios for which the potential outcomes significantly diverge from these comparable modes are those involving unconventional weapons, in which case the primary threat comes from the weapon rather than the delivery mode.

Given the availability of alternative attack modes and the uncertainties associated with the success of cruise missiles and UAVs to adversaries, we consider broad-based and expensive efforts focused only on this specific threat as appearing unrealistic. National, or even regional, efforts aimed at detection and warning and active defense require large investments that would either require significant increases in security funding or pull support away from defending against many of the alternative attack modes that are more readily available to adversaries than UAVs and cruise missiles. Unless significant additional benefits could be gained from such systems beyond simply addressing this threat, it would be difficult to justify their high costs.

Even though the level of threat is uncertain, nonproliferation efforts that seek to limit the availability of these systems are valuable. Given established efforts to maintain technology controls, we judge nonproliferation to be an element that is easy to include in a portfolio of defensive approaches at comparatively modest cost.

The uncertainties associated with the UAV/cruise-missile threat also make more general-purpose approaches (strategies appearing further to the upper right in Figure 5.2) appear particularly attractive. In fact, little, if any, new capability may be needed, since the United States is already pursuing these strategies in response to other threats. Investments in counterterrorism-intelligence activities and broader law-enforcement capability would increase the chances that a plot involving UAVs or cruise missiles would be discovered before it was executed, but would also produce other benefits even if this particular threat never appeared.[9,10] The additional cost of such investments would be modest relative to active-defense options.

Effective intelligence and law enforcement can significantly diminish the threat posed by terrorism; however, past experience has shown that some attack operations will inevitably evade even the best protective efforts. Response, recovery, and reconstitution capabilities represent general-purpose investments to address that eventuality, particularly because the payload capacities of many UAVs and cruise missiles mean that the scale of weapon they will deliver to a target will likely resemble that carried by more-common attack modes, such as pedestrian suicide bombers or moderate-scale vehicle bombs. As a result, the ability to bounce back from relevant UAV- and cruise missile–delivered attacks will also provide those capabilities for more familiar terrorist operational modes.

Investment in forensics capabilities appears attractive as well, since general capabilities will also provide benefits for other investigations that do not involve these particular weapons, and any specialized capabilities can be built at a national level to serve the country as a whole. The ability to gather evidence after an attack is critical both to breaking a campaign of such attacks and to aiding in attributing the source of technologies (e.g., if technologies can be traced back to states, additional diplomatic and other response modes could provide additional routes to respond to an attack, as well as reducing the potential willingness of states to share such technologies in the first place).

---

[9] For example, ongoing efforts focused on examining materiel entering the country to detect illicit cargoes (e.g., drugs, individuals entering the country illegally) could similarly watch for UAVs or cruise missiles being brought into the country; however, the challenges those efforts have encountered in keeping other threats outside the nation's borders underscore the difficulty inherent in these strategies.

[10] This statement assumes, of course, that the marginal benefits of adding additional resources to these activities would exceed their marginal cost—i.e., the United States has not "maxed out" what it can cost-effectively accomplish through these strategies. Recent discussions of both domestic-law-enforcement and foreign-intelligence activities suggest this is unlikely to currently be the case.

The intelligence community can make a significant contribution through its gathering of information on foreign UAVs. That information can be used to help inform law-enforcement agencies about what to look for in their day-to-day operations and to provide the basis for a detailed database that can be used for forensics. The Department of Defense and the Intelligence Community are uniquely situated to assist in this area because of their foreign-intelligence missions and dramatically greater resources than the law-enforcement community has at its disposal to gather detailed information on these systems. The key to gaining significant results from this type of activity is the ability to make intelligence available in an unclassified form (i.e., as law-enforcement-sensitive data) that can be shared widely with state and local law-enforcement organizations around the country. Such availability may limit the data largely to overtly procured systems, but even if the systems that are the subject of the most-sensitive foreign system–exploitation activities are unavailable, systems gathered by other means could prove quite useful.

Even with the best defensive efforts in place, UAV or cruise-missile attacks may occur in the future and lead to a demand to protect targets on an individual basis, rather than pursuing broad national defenses. If so, defense planners must make difficult decisions about how to deploy such selective defenses. From this perspective, our analysis of applications in which UAVs and cruise missiles might be attractive to adversaries provides little guidance. Of the five specific operational challenges we examined, the attractiveness of these systems in three of them—attacking across national borders, enabling campaigns, and carrying out many simultaneous attacks—has nothing to do with the specific target the adversary is planning to attack. Likewise, the large number of potential targets for an area attack with unconventional weapons is a theme that also provides little guidance on limiting potential sites for attack and, consequently, identifying deployment sites for point or small-area defenses. This possibility is similarly reinforced by the potential that a group's technology preferences or ready access to these technologies—the first of our two "planning paths" in Figure 4.1—may lead to their use on targets for which alternative attack modes might be comparable or even superior in effectiveness.

In an open and developed society, there will always be targets that are vulnerable from the air, so an adversary with one of these systems and the desire to use it will be able to find something to attack.

The only approach for focusing the deployment of active defenses, if they are demanded, must therefore be based not on potential adversary preferences but on those of the defense. Targets that are particularly vulnerable from the air might merit additional attention, since such vulnerability could provide an incentive for adversaries to go down this technological path. Targets that are of sufficient value and sensitive to attack by small numbers of moderate-payload systems may merit additional expenditure on protection. However, barring the observation of major shifts in the frequency of use of these systems by potential adversaries, protecting targets against these aerial threats should come only after those targets have defenses in place against simpler and potentially more-effective alternative attack modes. When examining UAVs and cruise missiles as modes for attacking over perimeters, these complex systems appeared attractive only when the defenses around the target were sufficiently robust that many alternatives were either not viable or had a high risk of failure.

Focusing on these threats over others could even produce adverse outcomes: If a target that could have been attacked with a truck bomb is instead attacked by an adversary using an explosives-laden UAV, the defense is actually better off: The truck could have delivered a much larger and deadlier payload than the aerial vehicle. If a defensive system to counter aerial attack led the attacker to instead use the truck, the country would be worse off for having the defenses in place. Strategies for such selective protection could include judiciously placed active defenses—although their cost would demand care in choosing how many targets were protected and would suggest that defensive systems that could be moved from place to place would have advantages as well. For targets judged to merit permanent protection, appropriately chosen passive defenses might be a more cost-effective choice to limit both the potential for efforts aimed at this threat and the competition for resources with other security needs, given that such defenses could also protect against a broader range of threats than just attack by UAVs and cruise missiles.

Deterring Asymmetrtic Use of UAVs and Cruise Missiles?

In considering defensive approaches to asymmetric threats from UAVs and cruise missiles, the preceding discussion has been largely silent on deterrence. *Deterrence* seeks to address the threat not by the direct effect of defensive and other measures, but by the ability of such measures to shape the decisionmaking of adversaries before an attack is staged. The ability to deter adversaries is frequently framed as coming from two mechanisms: *deterrence by punishment*, which seeks to change adversary behavior through threats of retribution if specific acts are carried out, and *deterrence by denial*, which seeks to change the apparent utility of carrying out those acts by implementing measures to prevent or degrade their effects.

Deterrence of a terrorist activity, whether that activity is participation in terrorist violence or a choice of specific attack modes or tactics, is not currently well understood, making the effects of any effort at deterrence difficult to anticipate.[11] However, even while this lack of knowledge might make deterrence a difficult element to build into an overall strategy aimed at this threat, it is worthwhile to consider the potential contribution that deterrence could make.

Deterrence by Punishment

In the literature on deterrence of terrorist activities, the effectiveness of deterrence by punishment has been called into question because of the level of commitment of many individual terrorists and organizations, particularly in view of the nature of the contemporary threat to the United States. Deterrence by punishment inherently requires an identifiable target to punish, and the clandestine nature of terrorist groups frequently makes identification difficult without significant effort.

Involvement of states could make this mode of deterrence more relevant, because a state implicated in such activities represents an identifiable future target for attack. Given the potential importance of states in providing terrorist groups with some types of these weapons, this

[11] For a review of this area, see Davis and Jenkins (2002); Stevenson (2004, pp. 179–185); National Research Council (2002); Casebeer and Thomas (2002); Melese and Angelis (2004, pp. 337–341); Carter (2001, pp. 84–102).

mode of deterrence could apply to limiting some uses of these technologies. Information for targeting punishment is an absolute necessity for implementing this strategy. Therefore, for the credibility of any such effort at deterrence, modes must be in place for gathering that information. Of the defensive approaches previously discussed, intelligence, law-enforcement, and forensic activities could contribute.

Deterrence by Denial

All the defensive approaches discussed could contribute to deterrence by denial because they could alter the effectiveness of any attack using UAVs and cruise missiles. Since even individual terrorists or terrorist organizations that are unconcerned by threats to their own safety or freedom seek to be successful in their operations, the increased risk that defenses create could provide disincentives to pursuing these attack modes.

Yet, even if this mode of deterrence does occur, its value is questionable. The level of defensive coverage needed to produce a deterrent away from these attack modes is unclear—and likely to differ from actor to actor. Full-coverage national systems certainly could provide such a deterrent, but at exceedingly high costs. Furthermore, even if deterrence away from UAVs and cruise missiles did occur, it might not actually reduce the threat of terrorism. As our analyses of alternative attack modes demonstrated, a wide variety of options exists for terrorists to attack targets in the United States. In some cases, these alternatives are potentially more destructive than UAVs and cruise missiles, making deterrence away from these modes negative from the perspective of minimizing the effects of an attack.

More-localized defenses could produce a deterrent for using these attack modes on specific defended targets, which could be valuable from the perspective of limiting attacks on particular targets of national significance—e.g., protecting major public events, key political targets (such as the State of the Union address), or highly critical infrastructure targets on which attacks could produce large, cascading damage and costs. However, the large number of targets that are otherwise vulnerable to these modes would presumably mean that the deterred threat would be displaced rather than eliminated. Consequently, whereas the "local" effect of deterrence by denial at and near the defended target would be considerable, there might not be any national reduction in the threat of terrorist attack or damage.

CHAPTER SIX

Conclusions

In this monograph, we examined the use of UAVs and cruise missiles for attacks within the United States. Whereas our examination of these systems from the adversary's point of view showed that they are viable options for a variety of attacks, they distinguish themselves from other potential options in only a few ways:

- Cruise missiles and UAVs stand out as an added threat to the few defended targets that currently exist in the United States.
- They enable the physical separation of the attack team from the site of an attack, which would allow an attacker to
  - carry out a campaign (a series of attacks over time)
  - stage multiple simultaneous attacks
  - initiate attacks from beyond the U.S. border.

In these circumstances, cruise missiles and UAVs may be attractive to an attacker. But, in most cases, alternative attack modes are similar or even superior.

UAVs and cruise missiles cannot be dismissed as potential threats, but they do not merit extensive specialized investments designed for mitigating them, particularly in a large nation in which a wide variety of potential targets is vulnerable to alternative attack modes.

Anticipating the Attractiveness of a Novel Threat to Adversaries

These conclusions were reached after conducting an analysis of alternatives from an adversary's perspective. When a novel threat is first recognized, it is easy to conclude that adversaries will be interested in pursuing it immediately. Given such a conclusion, it might be further concluded that fundamental changes are needed in defensive approaches to respond to the changed threat environment. Depending on the nature of the threat and the preferences of adversaries, both conclusions may be incorrect. By forcing comparison of a new threat to the capabilities already available to adversaries through alternative means, the analytic process described here can help inform conclusions about whether a new threat will be attractive to adversary groups and, if it is, the consequences for defense planning.

In this context, an analysis of alternatives seeks to anticipate the potential attractiveness of a novel tactic or technology by explicitly comparing it with alternative attack modes. Carrying out this analysis from the attacker's point of view helps to escape the trap of simply assuming that the same characteristics that make a technology attractive to a traditional military or security organization will necessarily make it attractive to a terrorist or other unconventional adversary.

Implications for the Defense

Even if an adversary chooses to use a new tactic or technology, what are the implications? From the defender's point of view, the primary concern is the outcome of adoption of a new attack mode by potential adversaries. In some cases, the effects of UAV or cruise- missile attacks are functionally indistinguishable from alternative modes. For example, a group could use a UAV or cruise missile to deliver explosives to an undefended target. When employed this way, the novel attack mode would not significantly increase the level of threat faced by the nation, since approaching an undefended target from the air would not significantly change the results of the attack. A similar explosive could be delivered to such a target by many other means.

The United States should design a defensive strategy built on broader defensive approaches that provide benefits not only against the threat posed by cruise missiles and UAVs but also for other terrorist and nonterrorist risks as well. Many activities are intended to uncover a terrorist attack before it occurs. Counterterrorist intelligence, law-enforcement activity, and forensics investigation all focus on identifying and catching the perpetrators either before or after an attack is carried out. Investments in counterterrorism and law enforcement will increase security against not only cruise missile and UAV attacks but also against all potential terrorist attacks.

A few investments are specific to cruise missiles and UAVs, such as gathering information to help law enforcement identify potential supply chains or conducting forensics analysis that could improve security at reasonable cost.

In the wake of an attack, forensic capability is needed to mine the attack scene for information and identify the perpetrators, potentially destroying their ability to carry out additional attacks. Such a capability could be particularly important for UAVs and cruise missiles, given that the separation of the attacker from the target distinguishes this threat from other attack options. Many forensic capabilities useful in this context would also contribute to other investigations, so investments in this area could produce benefits in areas unrelated to this threat. Specialized investments relevant to these threats also have a role to play. Collection of relevant technology and information to support the development of better forensics approaches—e.g., acquisition and study of foreign UAV and cruise-missile systems in ongoing efforts to gather and exploit technical intelligence—would play an important role in building the foundations needed for post-attack study and for determining any unique signatures of specific countries' systems. Systems for tracking aerial targets may be useful for forensic purposes, even if their performance is not good enough to support detection, warning, and active defense.

Through international agreements, the United States could also make it more difficult to get sophisticated cruise missiles and UAVs.

The MTCR, while certainly not a perfect instrument, is intended to constrain the transfer of missiles, including cruise missiles and UAVs with a range greater than 300 km, particularly those with payloads greater than 500 kg. Diplomatic efforts to strengthen the MTCR's transfer protocols focused on long-range, large-payload weapons could constrain the availability of some of the most destructive systems.

While seductive, large investments to defend against these air threats at the point of attack, particularly in the homeland, can distract from other more-productive defense investments that are focused on preventing attacks before they occur or in recovering evidence and performing forensics after an attack. Cruise-missile defenses would be costly; each system could defend only a small amount of territory, and even effective defensive performance within those areas would be exposed to operational challenges.

In an era in which a stated goal of U.S. adversaries is to damage the American economy and cause the United States to devote increasing amounts of its resources to defenses, ensuring that the government does not overspend to mitigate individual threats must be considered at the same time as trying to ensure that the nation is appropriately protected against terrorist and other asymmetric threats.

Bibliography

"Al Qaeda-Linked Group Takes Credit for Saudi Attack," *CNN*, December 7, 2004. As of April 21, 2007:
http://www.cnn.com/2004/WORLD/meast/12/06/jeddah.attack/

"Al-Qaeda Online: Understanding Jihadist Internet Infrastructure," *Jane's Intelligence Review*, January 1, 2006.

Baker, John C., Beth E. Lachman, David R. Frelinger, Kevin M. O'Connell, Alexander C. Hou, Michael S. Tseng, David T. Orletsky, and Charles W. Yost, *Mapping the Risks: Assessing the Homeland Security Implications of Publicly Available Geospatial Information*, Santa Monica, Calif.: RAND Corporation, MG-142-NGA, 2004. As of April 25, 2007:
http://www.rand.org/pubs/monographs/MG142/

"The Bali Bombing Plot," *BBC News*, October 12, 2002. As of February 3, 2007:
http://news.bbc.co.uk/2/hi/asia-pacific/3157478.stm

Barzilay, David, *The British Army in Ulster*, Volume 2, Belfast, Northern Ireland: Century Books, 1975.

———, *The British Army in Ulster*, Volume 4, Belfast, Northern Ireland: Century Books, 1981.

Bolkcom, Christopher, *Homeland Security: Defending U.S. Airspace*, Washington, D.C.: Congressional Research Service, RS21394, Updated June 6, 2006.

———, "Statement of Christopher Bolkcom, Analyst in National Defense Congressional Research Service, Before the Senate Governmental Affairs Committee, Subcommittee on International Security, Proliferation, and Federal Services," Hearing on Cruise Missile Proliferation, June 11, 2002.

Bowes, Peter, "High Hopes for Drone in LA Skies," *BBC News*, June 6, 2006. As of 24 April 2007:
http://news.bbc.co.uk/2/hi/americas/5051142.stm

Bureau of Alcohol, Tobacco, and Firearms, "ATF Vehicle Bomb Explosion Hazard and Evacuation Distance Tables," Washington, D.C.: ATF Instruction 5400.1, January 1999.

"Capitol Evacuated Before Reagan Procession," *CNN*, June 9, 2004. As of February 3, 2007:
http://edition.cnn.com/2004/ALLPOLITICS/06/09/capitol.evacuation/

Carter, Josh, "Transcending the Nuclear Framework: Deterrence and Compellence as Counter-Terrorism Strategies," *Low Intensity Conflict and Law Enforcement*, Vol. 10, No. 2, Summer 2001, pp. 84–102.

Casebeer, William, and Troy Thomas, "Deterring Violent Non-State Actors in the New Millennium," *Strategic Insights*, Vol. 1, No. 10, December 2002. As of April 24, 2007:
http://www.ccc.nps.navy.mil/si/dec02/terrorism2.asp

Cragin, Kim, and Sara A. Daly, *The Dynamic Terrorist Threat: An Assessment of Group Motivations and Capabilities in a Changing World*, Santa Monica, Calif.: RAND Corporation, MR-1782-AF, 2004. As of April 22, 2007:
http://www.rand.org/pubs/monograph_reports/MR1782/

Cragin, Kim, and Scott Gerwehr, *Dissuading Terror: Strategic Influence and the Struggle Against Terrorism*, Santa Monica, Calif.: RAND Corporation, MG-184-RC, 2005. As of April 24, 2007:
http://www.rand.org/pubs/monographs/MG184/

Daly, Sara A., John Parachini, and William Rosenau, *Aum Shinrikyo, Al Qaeda, and the Kinshasa Reactor: Implications of Three Case Studies for Combating Nuclear Terrorism*, Santa Monica, Calif.: RAND Corporation, DB-458-AF, 2005. As of April 25, 2007:
http://www.rand.org/pubs/documented_briefings/DB458/

David, Anthony, "Sri Lanka Rebels Bomb Fuel Depot, Closing Airport," *The New York Times*, April 30, 2007.

Davis, Paul, and Brian Michael Jenkins, *Deterrence and Influence in Counterterrorism: A Component in the War on al Qaeda*, Santa Monica, Calif.: RAND Corporation, MR-1619-DARPA, 2002. As of April 24, 2007:
http://www.rand.org/pubs/monograph_reports/MR1619/

Department of the Navy, Naval Historical Center, "War Chronology: January 1991," *U.S. Navy in Desert Shield/Desert Storm*, Washington, D.C., May 15, 1991. As of April 24, 2007:
www.history.navy.mil/wars/dstorm/dsjan2.htm

Dickerson, Larry, "UAVs on the Rise," *Aviation Week & Space Technology*, January 15, 2007, pp. 114–127.

FARSIGHT Products homepage.

FAS Intelligence Resource Program, *Report of the Accountability Review Boards on the Embassy Bombings in Nairobi and Dar es Salaam on August 7, 1998*, January 1999. As of April 21, 2007:
http://www.fas.org/irp/threat/arb/accountability_report.html

Federal Bureau of Investigation, *Uniform Crime Reports, 2002*. As of April 22, 2007:
http://www.fbi.gov/ucr/02cius.htm

Forecast International, *AGM-129 Advanced Cruise Missile*, archived February 2003. As of April 17, 2007:
http://www.forecastinternational.com/archive/mm/mm0109.doc

Gardner, Frank "Hezbollah Missile Threat Assessed," *BBC News*, August 3, 2006. As of August 22, 2006:
http://news.bbc.co.uk/2/hi/middle_east/5242566.stm

Geraghty, Tony, *The Irish War: The Hidden Conflict Between the IRA and British Intelligence*, Baltimore, Md.: The Johns Hopkins University Press, 2000.

Gips, Michael, "A Remote Threat," *Security Management*, October 2002.

Glasstone, Samuel, and Phillip J. Dolan, *The Effects of Nuclear Weapons*, 3rd ed., Washington, D.C.: U.S. Department of Defense and Energy Research and Development Agency, 1977.

Gormley, Dennis M., "Globalization and WMD Proliferation Networks: The Case of Unmanned Air Vehicles as Terrorist Weapons," *Strategic Insights*, Vol. 5, No. 6, July 2006.

———, "UAVs and Cruise Missiles as Possible Terrorist Weapons," in James Clay Moltz, ed., *New Challenges in Missile Proliferation, Missile Defense, and Space Security*, Monterey, Calif.: Monterey Institute of International Studies, Center for Nonproliferation Studies, Occasional Paper No. 12, July 2003, pp. 3–9.

———, "Unmanned Air Vehicles as Terror Weapons: Real or Imagined?" on International Institute for Counter-Terrorism Web site:
http://ict.org.il/apage/5285.php

Goshen-Meskin, Drora, "Presentation of the Eagle UAV System," Israel Aircraft Industries, Ltd., MALAT Division, Military Aircraft Group, July 11, 2005. As of May 3, 2007:
http://www.csl.ulg.ac.be/haas/WS1/11%20July%202005/Session%201/Presentation%204_Goshen.pdf

Harnden, Toby, *Bandit Country: The IRA and South Armagh*, London, United Kingdom: Coronet Books, LIR, 2000.

"The High-Flying UAV Marketplace (8/30/2004)," Military Periscope.com, 2004. As of May 3, 2007:
http://www.militaryperiscope.com/special/special-200408301146.shtml

Hilburn, Matt, "Hezbollah's Missile Surprise," *Today in the Military*, September 28, 2006. As of February 3, 2007:
http://www.military.com/forums/0,15240,115199,00.html

Hoffman, Bruce, "The Logic of Suicide Terrorism," *The Atlantic Monthly*, June 2003. As of August 22, 2006:
http://www.theatlantic.com/doc/200306/hoffman

House of Commons, *Report of the Official Account of the Bombings in London on 7th July 2005*, London, England, United Kingdom: The Stationery Office, May 11, 2006. As of August 22, 2006:
http://news.bbc.co.uk/1/shared/bsp/hi/pdfs/11_05_06_narrative.pdf

Jackson, Brian A., *Aptitude for Destruction*, Volume 1, *Organizational Learning in Terrorist Groups and Its Implications for Combating Terrorism*, Santa Monica, Calif.: RAND Corporation, MG-331-NIJ, 2005a. As of April 25, 2007:
http://www.rand.org/pubs/monographs/MG331/

———, "The Provisional Irish Republican Army," in Brian A. Jackson, John C. Baker, Kim Cragin, John Parachini, Horacio R. Trujillo, and Peter Chalk, *Aptitude for Destruction*, Volume 2, *Case Studies of Learning in Five Terrorist Organizations*, Santa Monica, California: RAND Corporation, MG-332-NIJ, 2005b. As of April 25, 2007:
http://www.rand.org/pubs/monographs/MG332/

———, "Technology Acquisition by Terrorist Groups: Threat Assessment Informed by Lessons from Private Sector Technology Adoption," *Studies in Conflict & Terrorism*, Vol. 24, May–June 2001, pp. 183–213.

Jackson, Brian A., John C. Baker, Kim Cragin, John Parachini, Horacio R. Trujillo, and Peter Chalk, *Aptitude for Destruction*, Volume 2, *Case Studies of Learning in Five Terrorist Organizations*, Santa Monica, Calif.: RAND Corporation, MG-332-NIJ, 2005. As of April 25, 2007:
http://www.rand.org/pubs/monographs/MG332/

Jackson, Brian A., Peter Chalk, Kim Cragin, Bruce Newsome, John Parachini, William Rosenau, Erin M. Simpson, Melanie Sisson, and Donald Temple, *Breaching the Fortress Wall: Understanding Terrorist Efforts to Overcome Defensive Technologies*, Santa Monica, Calif.: RAND Corporation, MG-481-DHS, 2007. As of April 25, 2007:
http://www.rand.org/pubs/monographs/MG481/

Jane's Terrorism and Insurgency Centre, "Exploding Toy Planes—The Next Threat to Israeli Security?" Coulsdon, Surrey, United Kingdom, February 24, 2003.

———, *The Madrid Rail Bombings*, Coulsdon, Surrey, United Kingdom: JTIC Terrorism Case Study No. 1, March 10, 2005.

Karmon, Ely, "Hizballah as Strategic Threat to Israel," on International Institute for Counter-Terrorism Web site:
http://ict.org.il/apage/5285.php

Lambakis, Steven, James Kiras, and Kristin Kolet, *Understanding "Asymmetric" Threats to the United States*, Fairfax, Va.: National Institute for Public Policy, September 2002. As of May 3, 2007:
http://missilethreat.com/repository/doclib/20021000-NIPP-asymmetricthreats.pdf

McGeer, T., and J. Vagners, "Wide-Scale Use of Long-Range Miniature Aerosondes Over the World's Oceans," Washington, D.C.: National Oceanic and Atmospheric Administration, 2000. As of April 24, 2007:
http://www.mmm.ucar.edu/uswrp/thorpex/observing/McGeer_Aerosonde.pdf

Melese, Francois, and Diana Angelis, "Deterring Terrorists from Using WMD: A Brinksmanship Strategy for the United Nations," *Defense and Security Analysis*, Vol. 20, No. 4, December 2004, pp. 337–341.

Miasnikov, Eugene, "Threat of Terrorism Using Unmanned Aerial Vehicles: Technical Aspects," Moscow, Russia: Center for Arms Control, Energy, and Environmental Studies, Moscow Institute of Physics and Technology, 2005.

Morin, Richard, and Claudia Deane, "Half of Area Residents in Fear, Post Poll Finds; Many Alter Routines as a Safety Measure," *The Washington Post*, October 24, 2002, p. A01.

Myre, Greg, "Israel Widens Scope of Attacks Across Lebanon," *The New York Times*, July 16, 2006.

National Intelligence Council, "Foreign Missile Developments and the Ballistic Missile Threat Through 2015," December 2001. As of April 22, 2007:
http://www.dni.gov/nic/PDF_GIF_otherprod/missilethreat2001.pdf

National Research Council, *Discouraging Terrorism: Some Implications of 9/11*, Washington, D.C.: National Academies Press, 2002.

Reuters News Service, "Bomb Plot Foiled in Colombia; Authorities Find Five Cars Packed with Explosives," *Houston Chronicle*, December 12, 2002, p. A31.

Richardson, Doug, "IDF Hunts Qassam-II Rocket Workshops," *Jane's Missiles and Rockets*, April 1, 2002.

———, "Qassam Rockets Tested in the West Bank," *Jane's Missiles and Rockets*, June 1, 2005.

Rojo, Aurelio, "The Security Model of Madrid Metro," Presentation at *Today's Realities for Rail Security*, American Public Transportation Association, New York, N.Y., June 11–14, 2006.

Stevenson, Jonathan, "Terrorism and Deterrence," *Survival*, Vol. 46, No. 4, November 2004, pp. 179–185.

Systems Assessment Group, NDIA Strike, Land Attack and Air Defense Committee, *Feasibility of Third World Advanced Ballistic Missile and Cruise Missile Threat*, Volume 2, *Emerging Cruise Missile Threat*, Arlington, Va.: National Defense Industry Association, August 1999.

"Troops Seize Rebels' Explosive Planes," Houston Chronicle News Services, August 27, 2002.

U.S. Department of Defense, Office of the Secretary of Defense, *Unmanned Aerial Systems Roadmap 2005–2030*, 2005. As of April 18, 2007:
http://www.acq.osd.mil/usd/Roadmap%20Final2.pdf

U.S. Government Accountability Office, *Precision-Guided Munitions: Acquisition Plans for the Joint Air-to-Surface Standoff Missile*, Washington, D.C.: Letter Report, GAO/NSIAD-96-144, June 28, 1996. As of April 18, 2007:
http://www.fas.org/man/gao/ns96144.htm

———, *Unmanned Aircraft Systems: Global Hawk Cost Increase Understated in Nunn-McCurdy Report*, Washington, D.C.: GAO-06-222R, December 15, 2005. As of April 17, 2007:
http://www.gao.gov/new.items/d06222r.pdf

U.S. Naval Institute, Periscope database. As of May 3, 2007:
http://www.militaryperiscope.com

U.S. Naval Institute, Periscope Web site:
http://www.militaryperiscope.com

Verton, Dan, "A View to a Kill: Terrorists and UAVs," *Homeland Defense Journal*, May 2005, pp. 10–14.

"Warheads," Introduction to Naval Weapons Engineering, ES310 Course Syllabus, 1998. As of April 20, 2007:
http://www.fas.org/man/dod-101/navy/docs/es310/warheads/Warheads.htm

Warnakulasuriya, Asanga, "Customs Seize Spy Planes," *Daily News*, November 20, 2003. As of February 8, 2007:
http://www.dailynews.lk/2003/11/20/new18.html

"Yamaha's RMAX—The Worlds Most Advanced Non-Military UAV," *Gizmag*, no date. As of April 17, 2007:
http://www.gizmag.com/go/2440/2/

Zaloga, Steven J., *UAVs: Interest Up*, Aviation Week Web site, no date. As of April 17, 2007:
http://www.aviationweek.com/aw/generic/story_generic.
jsp?channel=awstspace&id=news/sb03_10.xml